INTERNATIONAL
PHYSICS
OLYMPIADS

Volume 1

INTERNATIONAL

PHYSICS

OLYMPIADS

Volume 1

Editor: **Waldemar Gorzkowski**

Institute of Physics
Polish Academy of Sciences

World Scientific
Singapore • New Jersey • London • Hong Kong

Published by

World Scientific Publishing Co. Pte. Ltd.
5 Toh Tuck Link, Singapore 596224
USA office: 27 Warren Street, Suite 401-402, Hackensack, NJ 07601
UK office: 57 Shelton Street, Covent Garden, London WC2H 9HE

British Library Cataloguing-in-Publication Data
A catalogue record for this book is available from the British Library.

INTERNATIONAL PHYSICS OLYMPIADS

ISBN-13 978-981-02-0107-4
ISBN-10 981-02-0107-9

To My Mother
The Editor

FOREWORD

The International Physics Olympiad (IPhO) is an annual competition in physics for the secondary school students. In the spring of 1989 the Secretariat of the IPhO has circulated a letter to *all* participating countries. In the letter we asked our Colleagues and Friends from the International Board to send us reports on their national physics competitions and on the process of preparation of the participants for the international competition. We also asked for the most interesting physics problems that had been used in the national competitions or during the training of the teams before the international competition.

The aim of this inquiry was putting together the materials for this book. We have imagined that it might prove useful to the countries intend to initiate national physics competitions or to participate in the international event. It is also a means of exchange of experiences of the organizers from different countries. The important part of the book is the selection of problems quoted in the reports. The way of stating the problems and their scope differ somewhat in different countries. The awareness of these differences can prove both stimulating and useful.

This book is not intended solely for the readers involved in the physics competition movement. It should be equally appealing to the teachers, pupils and even younger university students who are interested in physics.

We have received seventeen reports in reply. All of them are presented in this book. We would like to express our gratitude to the authors for the preparation of camera-ready manuscripts, which made possible the direct reproduction of the articles to the exclusion of the complex, expensive and time consuming typesetting.

In addition to the aforementioned articles the book also contains the latest version of the Statutes of the IPhO, the Syllabus, the short history of the IPhO, the complete list of all the winners up to the present time and some statistical data. The latter are not to be treated as an official document of the IPhO but were prepared solely for the purpose of comparison.

We have also included an interesting analytical paper by B. Gau and R. Gau (GDR) concerning the assessment of the level of difficulty of theoretical problems presented at the competitions up to the present. We would like to thank the authors very warmly for their initiative and effort.

This book is only the first volume of what we hope to develop into a series. We hope that the "International Physics Olympiads" will become a semi-periodical. We would like to invite the cooperation of all those interested. Both criticism and suggestions should be directed to the Editor.

Finally I would like to thank my friends Marzena and Andrzej Reich for their unceasing and generous help in preparing the book for publication.

The Editor
Waldemar Gorzkowski
Instytut Fizyki PAN
al. Lotników 32/46
PL 00-668 Warszawa

CONTENTS

x

HISTORY

INTERNATIONAL PHYSICS OLYMPIADS HISTORY AND PERSPECTIVES

Waldemar Gorzkowski

Institute of Physics, Polish Academy of Sciences,
al. Lotników 32/46, 02-668 Warszawa, Poland

1. SHORT HISTORY

The International Physics Olympiad (IPhO) is an international physics competition for secondary school students. The first such competition was organized by Prof. Czesław Ścisłowski in Warsaw (Poland) in 1967. Since that time the International Physics Olympiads are organized, with few exceptions to be discussed later, in a different country every year.

The possibility of organizing the International Physics Olympiads was speculated prior to 1967. It was clear that the International Physics Olympiads should be an annual event like the International Mathematics Olympiad, already in existence, organized for the first time in 1959. The success of the International Mathematics Olympiads, and a positive experience gained from its organization, greatly stimulated physicists involved in physics education and interested in comparison of knowledge of the best students from different countries. The hard work and dedication of Professors: R. Kostial from Czechoslovakia, R. Kunfalvi from Hungary and Cz. Ścisłowski from Poland deserved particular praise. Each of them investigated various possibilities of organizing the first International Physics Olympiad in his country. It was concluded that Poland offered the best conditions and the most favorable atmosphere for such an event. This, together with a great personal contribution by Prof. Czesław Ścisłowski, resulted in the first international physics competition that took place in Warsaw in 1967.

One should underline here an essential difference between the International Mathematics Olympiads and the International Physics Olympiads: at the International Physics Olympiads the participants solve not only theoretical problems but also the experimental ones. For this reason the organization of the competition in physics is much more complicated and much more expensive.

Several months before the first IPhO took place, invitations were sent to all the East European countries. The invitations were accepted by Bulgaria, Czechoslovakia, Hungary and Romania (five countries including Poland, the organizer of the competition). Each team consisted of three secondary school students accompanied by one supervisor. The competition was arranged along the lines of the final stage of the Polish Physics Olympiad: one day for theoretical problems and one day for carrying out an experiment. One obvious difference was the participants having to wait for the scripts to be marked. During the waiting period the organizers arranged two excursions by plane to Kraków and to Gdańsk. At the first IPhO the students had to solve four theoretical and one experimental problems.

The second Olympiad was organized by Prof. R. Kunfalvi in Budapest, Hungary, in 1968. Eight countries took part in that competition – German Democratic Republic, Soviet Union and Yugoslavia joined the participating countries. Again, each country was represented by three secondary school students and one supervisor. Some time before the second IPhO a preliminary version of the Statutes and the Syllabus were produced. Later these documents were officially accepted by the International Board consisting of the supervisors of the teams that participated in the competition. This took place during a special meeting organized in Brno, Czechoslovakia, several months after the second IPhO. It is proper to underline that, in spite of various changes to be made later, all the basic features of the first Statutes remain valid to this day.

The third IPhO was arranged by Prof. R. Kostial in Brno, Czechoslovakia, in 1969. On that occasion each team consisted of five students and two supervisors. The competition in Brno was organized according to the official Statutes accepted earlier.

The next Olympiad took place in Moscow, Soviet Union, in 1970. Each country was represented by six students and two supervisors. During that Olympiad several small changes were introduced into the Statutes.

Since the fifth IPhO, held in Sofia, Bulgaria, in 1971, each team consists of five pupils and two supervisors.

The sixth IPhO was held in Bucharest, Romania, in 1972. It was an important event because among the participants there were present for the first time: the first out-of-European country (Cuba) and the first Western country (France). At this Olympiad the International Board decided to introduce several changes into the Statutes (however, no written proposal of the changes was produced).

Unfortunately, in 1973 there was no Olympiad because no

country was willing to organize it, although the number of participating countries exceeded the number of the past organizers. When it seemed likely that the International Physics Olympiads would die, Poland took the initiative of reviving the international competition and organized the VII IPhO in Warsaw in 1974 (for the second time). On this occasion the Federal Republic of Germany was invited to attend the competition for the first time. This fact certainly had a symbolic significance.

Before the competition, the Organizing Committee introduced into the Statutes the verbal changes discussed and accepted in Bucharest. The new version of the Statutes was sent to all the countries invited to the competition for acceptance or comments. The wording suggested by the Organizing Committee was accepted (with only one voice against). The most important changes were as follows:

a) the number of theoretical problems was reduced from four to three;

b) the number of working languages was reduced from four to two (English and Russian);

c) there should be one day of rest between the two examination days;

d) the criteria for prizes should be expressed in percentages with respect to the highest score received in given competition (formerly they had been counted with respect to the highest theoretically possible score).

In 1975, 1976 and 1977 the International Physics Olympiads took place in German Democratic Republic (for the first time), Hungary (for the second time), and Czechoslovakia (for the second time), respectively.

In spring 1977 in Ulan-Bator, Mongolia, there was a Conference of the Ministers of Education of Socialist Countries. The Conference decided that the socialist countries would organize the International Chemistry, Mathematics and Physics Olympiads every two years. This decision was a consequence of the increasing number of participating countries and rapidly increasing organizational costs. The above decision was commonly interpreted an implicit invitation to other countries to take charge of the international scientific olympiads. This explains why in 1978 and in 1980 there were no Olympiads (no non-socialist country was ready to organize the competition without a prior, necessary long-time preparation effort). The first IPhO organized by non-socialist country was the XIII IPhO that took place in Malente, FRG, in 1982. Then for the first time the participants solved two experimental problems (instead of one).

In 1983 the IPhO was organized, for the second time, in Bucharest, Romania. Here the number of problems prepared by the organizers for the pupils much exceeded the number of problems mentioned in the Statutes and the International Board spent a lot of time discussing the Statutes and the Syllabus and the future of the Olympiads.

As regards the future of the International Physics Olympiads, there was only one important decision made in Bucharest. It was decided that the next competition would take place in Sweden in 1984. Unfortunately, there were no volunteers to organize the Olympiads in 1985, 1986 and 1987. In such a situation, upon suggestion of Dr. Gunter Lind (FRG), the International Board decided to establish a permanent Secretariat (consisting of one person) for coordination of the long term work of the International

Physics Olympiads and for popularizing the Olympiads. At the same time it was decided that the Secretariat together with Prof. Lars Silverberg (Sweden), the organizer of the next competition in Sigtuna in 1984, should prepare a new version of the Statutes.

The project of the Statutes was completed and the new Statutes were accepted at the XV IPhO. There are, in fact, only minor differences between the old and new versions. The most essential difference is that the new version legalize the existence of the Secretariat of the International Physics Olympiads, consisting of two persons. Another change is that at the experimental part of the competition the participants can get one or two experimental tasks (earlier only one was allowed). One can say that the new version differs from the old one primarily in wording, which in the new version is much more precise.

The delegation heads (supervisors; two persons from each participating country) form the, so called, International Board, which is the highest authority of the International Physics Olympiads. The International Board does not change significantly from year to year. The majority of members know each other very well. In the International Board there is a very pleasant, friendly atmosphere. Thanks to this attitude and good will many difficult problems can be solved without great effort. This is why the Secretariat was able, for instance, to solve the problem of organization of the International Physics Olympiads in 1985, 1986 and 1987. In 1985 the International Physics Olympiad took place in Portorož (Yugoslavia), in 1986 – in London-Harrow (Great Britain) and in 1987 – in Jena (GDR).

Here we would like to emphasize that the United Kingdom organized the XVII IPhO in London—Harrow within only two years from its entry into the competition! It was made possible through hard work, great personal dedication and great enthusiasm of Dr. Cyril Isenberg, Dr. Guy Bagnall and Mr. William Jarvis.

Due to joint efforts of the Secretariat and the organizers of the competitions in 1985 (Prof. Anton Moljk and Dr. Bojan Golli) and in 1986 (Dr. Dr. Cyril Isenberg and Guy Bagnall) a new version of the Syllabus was produced. Its theoretical part was accepted in Portorož in 1985 and the practical one in London—Harrow in 1986. Quite recently, following a suggestion of the International Board, the Secretariat prepared a new, so called, "column version" of the Syllabus. This version shows not only the "breadth" of the physics contents but also the "depth" of approach required. The Syllabus of the International Physics Olympiads is indeed very modern. Nevertheless, the International Board is always ready to introduce improvements into the Statutes and Syllabus and does this if necessary.

Here we would like to point out that functioning of the Secretariat is so efficient owing to not only to personal efforts of its members but also to the kind assistance of the members of the International Board. It is proper to mention here the help of Dr. Gunter Lind (FRG), Prof. Helmuth Mayr (Austria), Frof. Lars Silverberg (Sweden), Mr. Nicola Velchev (Bulgaria) and many, many others.

It is obvious that the existence of the International Physics Olympiads itself is a result of certain international co-operation. More important is a long term

international co-operation between the members of the
International Board. This kind of co-operation exists since
the very beginning, i.e. since the first IPhO. The members
of the International Board exchange physics problems, books,
journals, articles, they discuss their experience gained
during organization of the national physics competitions
etc., etc. Due to such permanent, or semi-permanent,
contacts and due to existence of the International Physics
Olympiads some countries have organized national physics
olympiads or, at least, smaller scale competitions for
selecting the teams to attend the international
competition. Many countries have improved their national
syllabuses on physics by introducing new approaches (e.g. in
thermodynamics), new topics (e.g. relativity, quantum
physics), or by reducing some parts of too traditional
character (e.g. geometric optics).

The significance of the International Physics Olympiads
is continually increasing. The role of the International
Physics Olympiads is recognized also by such international
organizations as UNESCO and the EPS (European Physical
Society).

The first contacts with UNESCO took place way back, in
1968, but more extensive co-operation began in 1984. In the
period 1984 - 1988 UNESCO supported financially the
publication of the proceedings of the of the subsequent
Olympiads. The proceedings were distributed to all the
countries-members of UNESCO. It gave us favorable publicity.
In addition, UNESCO has published several books on the
physics Olympiads in various languages. Quite recently, in
March 1989 UNESCO organized a meeting in Holland on the
"Future Development on Mathematics and Science Olympiads".
The conclusions of this meeting are included in this issue.

The help of UNESCO is very valuable. One should **realize**, however, that the purposes of UNESCO are not identical with the purposes of the International Physics Olympiads (although many points are common). For example, too quick, artificially forced, increase in the number of participating countries can cause very serious organizational (mainly financial) problems. Simply, the International Physics Olympiad should be self-governing organization and its policy should be decided by the International Board only.

The co-operation with the EPS is slightly different as the financial possibilities of the EPS are not too great. The EPS gives us very strong moral support as well as favorable publicity and propagates our achievements among the countries-members of the EPS. It was the EPS that inspired us in preparation and publication of the booklet entitled "Procedures for Selecting Teams to the International Physics Olympiads". The booklet comprises a compilation of reports of different delegations and is very important and helpful for the countries wishing to join the competition. The booklet was prepared by the Secretariat together with Prof. Lars Silverberg and published by him privately in Lund (Sweden). In 1989 the EPS created a permanent special prize for the winner of the Olympiad for reaching the best equilibrium between the theoretical and experimental parts of the competition.

The Tables I and II present some data concerning the past and future International Physics Olympiads.

2. COMPETITION

The competition lasts for two days. The first day is devoted to theory (three problems involving at least four areas of physics taught in secondary schools). Another day is devoted to experiment (one or two problems). These two days are separated by at least one day of rest. The time allotted for solving the problems is five hours (on both occasions). Each team consists of students from general or technical secondary schools (not colleges or universities). Typically each team consists of five students (pupils) and two supervisors. The latter form the International Board. For further details concerning the teams see the Statutes of the International Physics Olympiads included in this issue.

The competition problems are prepared by the organizers of the Olympiad, i.e. by the host country. The texts of the problems and their solutions are analyzed in detail during the sessions of the International Board. The International Board accepts the final wording of the problems in two working languages, i.e. in English and Russian. The texts of the German and French versions are also provided, but the wording in these languages is not analyzed. In addition, the International Board suggests an evaluation guideline or accepts suggestions of the organizers concerning this matter. The theoretical and experimental problems are discussed separately, on the day prior to each part of the competition.

When the final written versions of the problems are ready, the members of the International Board translate them into mother tongue(s) of the students. Thus, each student receives the problems in his/her language (and writes the

13

Table I

COUNTRIES PARTICIPATING IN THE INTERNATIONAL PHYSICS OLYMPIADS

```
Olympiad                          1...                    2...
                1 2 3 4 5 6 7 8 9 0 1 2 3 4 5 6 7 8 9 0

Year            196...197...          198...
                7 8 9 0 1 2 4 5 6 7 9 1 2 3 4 5 6 7 8 9
```

#	Country	1	2	3	4	5	6	7	8	9	0	1	2	3	4	5	6	7	8	9	0
1.	Australia	-	-	-	-	-	-	-	-	-	-	-	-	-	-	-	-	O	#	#	#
2.	Austria	-	-	-	-	-	-	-	-	-	-	-	-	#	#	#	#	#	#	H	#
3.	Belgium	-	-	-	-	-	-	-	-	-	-	-	-	-	-	-	-	-	O	#	#
4.	Bulgaria	#	#	#	#	H	#	#	#	#	#	#	H	#	#	#	#	#	#	#	#
5.	Canada	-	-	-	-	-	-	-	-	-	-	-	-	-	-	-	O	#	#	#	#
6.	China	-	-	-	-	-	-	-	-	-	-	-	-	-	-	O	O	#	#	#	#
7.	Colombia	-	-	-	-	-	-	-	-	-	-	-	-	-	-	-	-	-	O	#	#
8.	Cuba	-	-	-	-	-	#	-	-	-	-	-	-	-	#	#	#	#	#	#	#
9.	Cyprus	-	-	-	-	-	-	-	-	-	-	-	-	-	-	-	-	-	-	#	#
10.	Czechoslovakia	#	#	H	#	#	#	#	#	#	#	H	#	#	#	#	#	#	#	#	#
11.	Denmark	-	-	-	-	-	-	-	-	-	-	-	-	-	-	-	-	-	-	O	-
12.	Finland	-	-	-	-	-	-	-	-	O	#	#	#	#	#	#	#	#	#	#	#
13.	France	-	-	-	-	-	#	-	#	#	#	-	#	#	#	-	-	-	-	-	-
14.	FRG	-	-	-	-	-	-	#	#	#	#	#	#	H	#	#	#	#	#	#	#
15.	GDR	-	#	#	#	#	#	#	#	H	#	#	#	#	#	#	#	#	H	#	#
16.	Great Britain	-	-	-	-	-	-	-	-	-	-	-	-	-	O	#	#	H	#	#	#
17.	Greece	-	-	-	-	-	-	-	-	-	-	-	-	#	-	-	-	-	O	-	O
18.	Hungary	#	H	#	#	#	#	#	#	#	H	#	#	#	#	#	#	#	#	#	#
19.	Iceland	-	-	-	-	-	-	-	-	-	-	-	-	-	-	#	#	#	#	#	#
20.	Iran	-	-	-	-	-	-	-	-	-	-	-	-	-	-	-	-	-	-	O	#
21.	Italy	-	-	-	-	-	-	-	-	-	-	-	#	#	-	-	O	-	#	#	#
22.	Kuwait	-	-	-	-	-	-	-	-	-	-	-	-	-	-	-	-	O	#	#	#
23.	Lithuanian SSR	-	-	-	-	-	-	-	-	-	-	-	-	-	-	-	-	-	-	-	U
24.	Netherlands	-	-	-	-	-	-	-	-	-	-	-	-	#	#	#	#	#	#	#	#
25.	Norway	-	-	-	-	-	-	-	-	-	-	-	-	-	-	#	#	#	#	#	#
26.	Poland	H	#	#	#	#	#	H	#	#	#	#	#	#	#	#	#	#	#	#	H
27.	Romania	#	#	#	#	#	H	#	#	#	#	#	#	#	H	#	#	#	#	#	#
28.	Singapore	-	-	-	-	-	-	-	-	-	-	-	-	-	-	-	-	-	-	W	#
29.	Soviet Union	-	#	#	H	#	#	#	#	#	#	H	#	#	#	#	#	#	#	#	#
30.	Spain	-	-	-	-	-	-	-	-	-	-	-	-	-	-	-	-	-	-	O	O
31.	Sweden	-	-	-	-	-	O	-	-	#	#	#	#	#	#	H	#	#	#	#	#
32.	Thailand	-	-	-	-	-	-	-	-	-	-	-	-	-	-	-	-	-	-	-	O
33.	Turkey	-	-	-	-	-	-	-	-	-	-	-	-	-	-	-	#	#	#	-	#
34.	UAE	-	-	-	-	-	-	-	-	-	-	-	-	-	-	-	-	-	-	-	O
35.	USA	-	-	-	-	-	-	-	-	-	-	-	-	-	-	-	O	#	#	#	#
36.	Vietnam	-	-	-	-	-	-	-	-	-	-	-	#	#	#	#	#	-	#	#	-
37.	Yugoslavia	-	#	#	#	-	-	-	-	-	#	#	#	#	#	#	H	#	#	#	#
	UNESCO	-	O	-	-	-	-	-	-	-	-	-	O	-	-	-	-	O	O	O	-
	EPS	-	-	-	-	-	-	-	-	-	-	-	-	-	-	O	O	O	O	O	O

→

14

Explanation to the Table I:

: participation   − : no participation
H : host country   O : observer
W : willingness to start from next year declared
U : unofficial participiation (guest of the organizers)

Table II

 O R G A N I Z E R S   O F   T H E
I N T E R N A T I O N A L   P H Y S I C S   O L Y M P I A D S

Past:

I	1967	**Warsaw** (Poland)
II	1968	**Budapest** (Hungary)
III	1969	**Brno** (Czechoslovakia)
IV	1970	**Moscow** (Soviet Union)
V	1971	**Sofia** (Bulgaria)
VI	1972	**Bucharest** (Romania)
VII	1974	**Warsaw** (Poland)
VIII	1975	**Guestrow** (GDR)
IX	1976	**Budapest** (Hungary)
X	1977	**Hradec Kralove** (Czechoslovakia)
XI	1979	**Moscow** (Soviet Union)
XII	1981	**Varna** (Bulgaria)
XIII	1982	**Malente** (FRG)
XIV	1983	**Bucharest** (Romania)
XV	1984	**Sigtuna** (Sweden)
XVI	1985	**Portoroż** (Yugoslavia)
XVII	1986	**London-Harrow** (United Kingdom)
XVIII	1987	**Jena** (GDR)
XIX	1988	**Bad Ischl** (Austria)
XX	1989	**Warsaw** (Poland)

Future:

XXI	1990	**Groningen** (The Netherlands)
XXII	1991	**Havana** (Cuba)
XXIII	1992	**Helsinki** (Finland)
XXIV	1993	? (USA) − not confirmed
XXV	1994	? (China) − not confirmed
XXVI	1995	? (Australia) − not confirmed
XVII	1996	? (Norway)

solutions likewise).

The assessment of the pupils' entries is performed by the Organizing Committee, with the help of local interpreters, in accordance with the general evaluation guidelines accepted earlier by the International Board. Later the grading is agreed with the delegation leaders (supervisors) of each team. The final results are accepted by the International Board.

The maximum number of points available for each problem is allocated by the organizers, but the ratio of the total number of points for the theoretical problems to the total number of points for the experimental problem(s) should be 3 : 2. In practice, the maximum number of points available for each theoretical problem is 10 and for each experimental problem 10 or 20 (10 when there are two experimental problems, 20 when there is only one experimental problem).

The winners of the competition receive diplomas/medals or honorable mentions according to the following rules:

The mean value of points accumulated by the three best participants at the given contest is considered as 100%.
The contestants who obtain more than 90% of the above mentioned mean value receive first prizes. The contestants who obtain between 78% and 90% receive second prizes. The contestants who obtain between 65% and 78% receive third prizes. The contestants who obtained between 50% and 65% receive honorable mentions.
All other participants receive certificates of participation in the competition. The participant with the highest score (Absolute Winner) receives an additional prize. Some special prizes can also be awarded.

We would like to underline that the number of the first prizes is not limited. The same refers to each other category of prizes. Therefore, the increase in the score of some participants (following, for example, a discussion between the supervisors and the markers) from the group of second prize winners to the group of first prize winners does not change category of prize of any other participant. Thus, the delegation leaders do not compete against each other.

One can ask: what about a team classification? The answer is very simple: such a classification does not exist. The IPhO is a competition between individuals and the Statutes establish no way of defining a team result. Nevertheless, some people try to establish a kind of unofficial team classification. Some of them take a direct sum of scores as the result of the team. Some of them take the sum of scores of the three best participants in each team. Some of them take, for each team, the tree best results in each problem independently and so on, and so on. Of course, the final table depends on the way used when calculating the team results, and probably you can always find some strange system of counting the team result that will show your team to be the best or, at least, one of the best.

The assessment and its verification takes several days. Throughout this time the organizers and the International Board are hard pressed. The pupils, however, are free and this is a very pleasant time for them, real rest with various cultural events, plays, excursions, etc.

At the end of the competition the results are announced

at an official closing ceremony at which representatives of the Ministry of Education, representatives of various scientific institutions, famous physicists, etc. are present.

3. PERSPECTIVES

The financial principles of the organization of the competition are the following:

*** the country which sends the team covers the return travel costs (to and from the place of the competition) of the pupils and the accompanying persons;

*** from the moment of arrival until the moment of departure all the costs are covered by the organizing country. In particular, this concerns the costs of local travels, lodging, excursions, awards, etc.

The number of participating countries is continually increasing. As a result, the organization of the competition becomes more and more expensive. Moreover, it is more and more difficult to organize the experimental part of the competition so that all the students have the same experimental conditions of work.

We can ask, what will be the maximum number of countries? How long can the number of participating countries increase without any perturbations (assuming the same structure of the competition)? Should we start thinking about "Olympic Villages"?

Until now the organizers were always able to solve all the organizational problems related to the increasing number of participants. In our opinion the maximum number of countries present at a given Olympiad should not exceed sixty. Sixty countries times five students from each country makes 300 experimental stands. This is a very great number. Some countries, however, are able to provide such a number of identical experimental stands. Other countries can organize the experimental problem in two groups. It seems that no country will be able to organize the competition (in its present form) for more than sixty countries.

Can this number be reached? Theoretically, yes. But practically, probably not. The travel expenses can limit the number of participants. Many countries may not be able to send their teams to the competition every year. The number of participating countries will probably oscillate around fifty, depending on where the organizing country is situated. This will not require "Olympic Villages".

The financial situation of the of the organizers depends on the internal situation in the organizing country. In some countries all or almost all the costs are paid by the Ministry of Education. In such case the financial situation of the organizers is easy, much easier than in the case of countries in which the IPhO is treated as a private event. In the last case the budget consists of funds received from many different companies and depends very strongly on the general activity of the Organizing Committee. Rich companies are always able to support the organizers, but additional effort is required from the people involved in the organization of the competition.

In our opinion, it is not necessary to change anything in the structure of the International Physics Olympiads, at least at present when the number of participating countries is 30. We believe that even in the future all the financial and organizational difficulties will be successfully solved and no artificial barriers should be set up against the free increase of the number of participants.

Another important point, concerning the future, refers to the competition problems, their difficulty, creative capability necessary to solve them, etc. It seems that the problems given at several recent Olympiads were not as challenging as the problems presented years earlier. The creative capability necessary for solving is declining. They are too difficult, the numerical and mathematical calculations are too complicated.

It is not easy to explain this decline. Probably, the most important reason is that the number of participants is increasing. As a consequence, the organizers try to formulate the problems so that the marking is as simple as possible. For example, sometimes the exact way of solving is indicated in the text of the problem and the pupils should merely follow it step by step without too great great mental effort. Second example: solutions to some problems are associated with tedious mathematical calculations, alien to the true nature of physics, with strong chances of errors. Another reason is, probably, that the organizers try to include too many ideas and concepts in the texts of the problems, e.g. too many modern developments of physics, too many national achievements and so on. As a result, the texts of the problems are very long. Sometimes it is impossible to grasp the essence of the problem in a reasonable time.

The next point concerns the computer problems. One should realize that the computer problems, especially when a computer company is one of the main sponsors, cannot be avoided. However, we should always remember that the computer in only a kind of instrument. Nothing more! The

computer experiments, for example, are never real
experiments. The Brownian motion observed through a
microscope is a real phenomenon. But the Brownian motion
produced by the computer and observed on the screen is not –
it is a simulation of a real phenomenon only. We are in
favour of using computers at the Olympiad, but we should be
careful about their applications. Computers should not
dominate the physics contents of the problem. Moreover, we
must not allow the students from one country to have an
advantage over those from other countries.

The above critical remarks about the competition
problems of the several recent Olympiads stem from
comparison between the earlier and the more recent years. In
order to avoid any misunderstanding, we would emphasize here
that in an absolute scale both the earlier and the later
competition problems are of a high standard, they are
stimulating and require great creativity from the
participants. Some criticism, however, is necessary. It can
be useful in future improvements of our Olympiads.
 The last point we would like to mention here is the problem
of translations. This problem has several aspects.

The first of them concerns various printed materials
prepared by the Organizing Committee in a written form.
Usually this work is carried out very well and no
improvements are necessary.

The next aspect concerns the oral translations at the
sessions of the International Board. As it was mentioned
earlier, there are only two working languages: English and
Russian. Thus, there is only one translation channel:
English – Russian. We should say that, in general, the
translations at the sessions are not satisfactory and must

be improved on. It is not easy, however, to find an interpreter combining good knowledge of both the languages, good knowledge of physics and ability for simultaneous translation from English into Russian and vice versa. Moreover, it is very expensive. Any help from international organizations in this connection would be gratefully accepted by the organizers. Unfortunately, by now no international organization, including UNESCO, showed signs of willingness to help us on this matter.

The subsequent aspect concerns the translation of the pupils' solutions to the problems. The marking of the solutions is performed by the Organizing Committee which is responsible for correct translation. In the case of languages spoken by a number of countries, such as German or Spanish, there are no serious difficulties. Also there are no difficulties in the case of nations or countries with a great diaspora (e.g. Poland). But in the case of certain fringe languages the organizers sometimes face great problems. Fortunately, all the possible mistakes made during the marking procedure can be corrected at the verification sessions with the participation of the delegation leaders, although sometimes it is time consuming.

The last aspect refers to the pupils' interpreters to and from their mother tongues during excursions, plays and various social events. This aspect is of less importance as all the meetings organized for the pupils have a very relaxed, friendly atmosphere and the quality of translations is not emphasized. It may even be said, although it sounds a little odd, that bad translation is sometimes desirable since then the participants try to speak in foreign languages. Such a linguistic training is necessary and useful for everybody.

4. LITERATURE

[1] Nicolaus Vermes, **International Physics Competitions 1967 - 1977,** Roland Eötvös Physical Society, Budapest 1978, (first edition in Hungarian, second edition in English)

[2] K. K. Kudawa, **International Physics Olympiads.** Ganatleba, Tbilisi 1983 (in Georgian)

[3] **Proceedings of the 15. International Physics Olympiad, Sigtuna, Sweden,** AVC, Lund 1984 (sponsored by UNESCO)

[4] R. Kunfalvi, **Collection of Competition Tasks from the I through XV International Physics Olympiads 1967 - 1984,** Roland Eötvös Physical Society and UNESCO, Budapest 1985

[5] O. F. Kabardin, V. A. Orlov, **International Physics Olympiads for Pupils** (in Russian), Nauka, Moskva 1985

[6] **16th International Physics Olympiad, Portoroz, Yugoslavia,** ed. by A. Moljk and B. Golli, Society of Mathematicians, Physicists and Astronomers of Slovenia, Ljubljana 1985 (sponsored by UNESCO)

[7] Gunter Lind, **Physikalische Olympiade-Aufgaben.** Aulis Verlag Deubner & Co., Köln 1986 (in German)

[8] **17th International Physics Olympiad, Harrow-London, England,** Organizing Committee of the 17th IPhO. Harrow 1986 (sponsored by UNESCO)

[9] **18th International Physics Olympiad Report,** Ministry of Education of the GDR and Organizing Committee of the 18th IPhO. Eggersdorf 1987 (sponsored by UNESCO)

[10] **Collection of Competition Tasks from the I through XVII International Physics Olympiads** (in Chinese), published in China under aegis of UNESCO, Beijing 1988

[11] **Recueil des Sujets de Concours des Olympiades Internationales de Physique (1967 - 1984) et de**

Mathematique (1978 -1985), D. CROS - C.I.F.E.C.-UNESCO, Paris 1988

[12] **Procedures for Selecting Teams to the International Physics Olympiads** (a compilation of reports from different delegations), ed. by W. Gorzkowski, A. Kotlicki, L. Silverberg, published by L. Silverberg, Lund 19886.

[13] **Olimpiadas Internacionales de Fisica 1967 - 1986,** UNESCO and Oficina Regional de Educacion para America Latina y El Caribe, Santiago de Chile 1988

[14] **19th International Physics Olympiad - Report,** Federal Ministry of Education, Arts and Sports of the Republic of Austria, Vienna 1988 (sponsored by UNESCO)

[15] **Olimpiadas Internacionales de Fisica - I a XV (1967 - 1984),** translated by Teresa Martin Sanchez and Manuela Martin Sanchez, Instituto de Ciencias de la Educacion and Ediciones Universidad de Salamanca, Salamanca 1989

[16] **XX International Physics Olympiad,** ed. by W. Gorzkowski, World Scientific Publishing Company, Singapore 1990

[17] **International Physics Olympiads, vol. I** (collection of national reports), ed. by W. Gorzkowski, World Scientific Publishing Company, Singapore - in print

[18] W. Gorzkowski, A. Kotlicki, **Międzynarodowe olimpiady fizyczne,** WSiP, Warszawa - in print (in Polish)

RULES

S T A T U T E S
O F   T H E
I N T E R N A T I O N A L   P H Y S I C S   O L Y M P I A D S

(Adopted in Sigtuna, Sweden, June 1984;
changes: Bad Ischl, Austria, June 1988;
Warsaw, Poland, July 1989)

§ 1

In recognition of the growing significance of physics in all fields of science and technology, and in the general education of young people, and with the aim of enhancing the development of international contacts in the field of school education in physics, an annual physics competition has been organized for secondary school students, the competition is called the 'International Physics Olympiad' and is a competition between individuals.

§ 2

The competition is organized by the Education Ministry or another appropriate institution of one of the participating countries on whose territory the competition is to be conducted. Hereunder, the term 'Education Ministry' is used in the above meaning. The organizing country is obliged to ensure equal participation of all the delegations, and to invite all the participants of any of the last three competitions. Additionally, it has the right to invite other countries.

Within five years of its entry in the competition a country should declare its intention to be the host for a future Olympiad. This declaration should propose a timetable so that a provisional list of the order of countries willing to arrange Olympiads can be compiled.

A country which refuses to organize the competition may be barred from participation, even if delegation from that country has taken part in previous competitions.

§ 3

The Education Ministries of the participating countries, as a rule, assign the organization, preparation and execution of the competition to a physics society or another institution in the organizing country. The Education Ministry of the organizing country notifies the Education Ministries of the participating countries of the name and address of the institution assigned to the organization of the competition.

§ 4

Each participating country sends a team consisting of students of general or technical secondary schools, i.e. schools which cannot be considered technical colleges. Also students who finished their school examination in the year of the competition can be members of a team as long as they do not start the university studies. The age of the participants should not exceed twenty on June 30th of the year of the competition. Each team should normally have 5 members.

In addition to the students, two accompanying persons are invited from each country, one of whom is designated delegation head (responsible for whole delegation), and the other - pedagogical leader (responsible for the students). The accompanying persons become members of the International Board, where they have equal rights.

The delegation head and pedagogical leader must be selected from specialists in physics or physics teachers, capable of solving the problems of the competition competently. Normally each of them should be able to speak one of the working languages of the International Physics Olympiads.

The delegation head of each participating team should, on arrival, hand over to the organizers a list containing personal data on the contestants (surname, name, date of birth, home address, type and address of the school attended).

§ 5

The working languages of the International Physics Olympiad are English and Russian. Problems and solutions have also to be translated into German and French.

§ 6

The financial principles of the organization of the competition are as follows:

* The Ministry which sends the students to the competition covers the return travel costs of the students and the accompanying persons to the place at which the competition is held.

* All other costs from the moment of arrival until the moment of departure are covered by the Ministry of the organizing country. In particular, this concerns the costs for board and lodging for the students and the accompanying persons, the cost of excursions, awards for the winners, etc.

§ 7

The competition is conducted on two days, one for the theoretical competition and one for the experimental competition. There should be at least one day of rest between these two days. The time allotted for solving the problems should normally be five hours The number of theoretical problems should be three and the number of experimental problems one or two.

When solving the problems the contestants may make use of tables of logarithms, tables of physical constants, slide-rules, non-programmable pocket calculators and drawing material. These aids will be brought by the students themselves. Collections of formulae from mathematics or physics are not allowed.

The theoretical problems should involve at least four areas of physics taught at secondary school level (see Appendix). Secondary-school students should be able to solve the competition problems with standard high school mathematics and without extensive numerical calculation.

§ 8

The competition tasks are chosen and prepared by the host country.

§ 9

The marks available for each problem are defined by the organizer of the competition, but the total number of points for the theoretical problems should be 30 and for the experimental 20. The laboratory problems should consist of theoretical analysis (plan and discussion) and experimental execution.

The winners will receive diplomas or honourable mentions in accordance with the number of points accumulated as follows:

The mean number of points accumulated by the three best participants is considered as 100%

The contestants who accumulate more than 90% of points receive first prize (diploma).

The contestants who accumulate more then 78% up to 89% receive second prize (diploma).

The contestants who accumulate more than 65% up to 77% receive third prize (diploma).

The contestants who accumulate more than 50% up to 64% receive an honourable mention

The contestants who accumulate less than 50% of points receive certificates of participation in the competition.

The mentioned marks corresponding to 90%, 78%, 65% and 50% should be calculated by rounding off to the nearest lower integers.

The participant who obtains the highest score will receive a special prize and diploma.

Special prizes can be awarded.

§ 10

The obligations of the organizer:

a) The organizer is obliged to ensure that the competition is conducted in accordance with the Statutes.

b) The organizer should produce a set of 'Organization Rules', based on the Statutes, and send them to the participating countries in good time. These Organization Rules shall give details of the Olympiad not covered in the Statutes, and give names and addresses of the institutions and persons responsible for the Olympiad.

c) The organizer establishes a precise program for the competition (schedule for the contestants and the accompanying persons, program of excursions, etc.), which is send to the participating countries in advance.

d) The organizer should check immediately after the arrival of each delegation whether its contestants meet the conditions of the competitions.

e) The organizer chooses (according to § 7 and the list of physics contents in the Appendix to these Statutes) the problems and ensures the translation of the chosen problems and their solutions into the languages set out in § 5. It is advisable to select problems where the solutions require a certain creative capability and a considerable level of knowledge. Everyone taking part in the preparation of the competition problems is obliged to preserve complete secrecy.

f) The organizer must provide the teams with interpreters.

g) The organizer must supply interpreters for the working languages who are to be available at the sessions of the International Board. The interpreters should be able to cope with physical terminology.

h) The organizer should provide the delegation leaders with photostat copies of the solutions of the contestants in their delegation before the final classification.

i) The organizer is responsible for the grading of the problem solutions.

k) The organizer drafts a list of participants proposed as winners of the prizes and honourable mentions.

l) The organizer prepares the prizes (diplomas), honourable mentions and awards for the winners of the competition.

§ 11

The scientific part of the competition must be within the competence of the International Board, which includes the delegation heads and pedagogical leaders of all the delegations.

The Board is chaired by a representative of the organizing country. He is responsible for the preparation of the competition and serves on the Board in addition to the accompanying persons of the respective teams.

Decisions are passed by a majority vote. In the case of equal number of votes for and against, the chairman has the casting vote.

§ 12

The delegation leaders are responsible for the proper translation of the problems from the languages mentioned in § 5 to the mother tongue of the participants.

§ 13

The International Board has the following responsibilities:

a) to direct the competition and supervise that it is conducted according to the regulations;

b) to ascertain, after the arrival of the competing teams, that all their members meet the requirements of the competition in all aspects. The Board will disqualify those contestants who do not meet the stipulated conditions. The costs incurred by a disqualified contestant are covered by his country;.

c) to discuss the Organizers' choice of tasks, their solutions and the suggested evaluation guidelines before each part of the competition. The Board is authorized to change or reject suggested tasks but not to propose new ones. Changes may not affect experimental equipment. There will be a final decision on the formulation of tasks and on the evaluation guidelines. The participants in the meeting of the International Board are bound to preserve secrecy concerning the tasks and to be of no assistance to any of the participants;

d) to ensure correct and just classification of the prize winners;

e) to establish the winners of the competition and make a decision concerning presentation of the prizes and honourable mentions. The decision of the International Board is final;

f) to review the results of the competition.

g) to select the country which will be assigned the organization of the next competition.

Observers may be present at the meetings of the International Board, but not to vote or take part in the discussion.

§ 14

The institution in charge of the Olympiad announces the results and presents the awards and diplomas to the winners at an official gala ceremony. It invites representatives of the organizing Ministry and scientific institutions to the closing ceremony of the competition.

§ 15

The long term work involved in organizing the Olympiads is coordinated by a 'Secretariat for the International Physics Olympiads'. This Secretariat consists of a Secretary and Vice-Secretary normally from the same country. They are elected by the International Board for a period of five years when the chairs become vacant.

§ 16

The present Statutes have bees drafted on the basis of experience gained during past international competitions.

Changes in these Statutes, the insertion of new paragraphs or exclusion of old ones, can only be made by the International Board and requires qualified majority (2/3 of the votes).

No changes may be made to these Statutes or Syllabus

unless each delegation obtained written text of the proposal at least three months in advance.

§ 17

Participation in an International Physics Olympiad signifies acceptance of the present Statutes by the Education Ministry of the participating country.

§ 18

The originals of these Statutes are written in English and Russian.

Appendix to the Statutes of the
International Physics Olympiads

THE SYLLABUS

(The column version of the 1985/86 Syllabus with slight modifications, preliminarily accepted in Warsaw, Poland, in July 1989. To be discussed and accepted in the final form in Groningen, The Netherlands, in July 1990)

General

Adopted in Portorož, Yugoslavia, June 1985

a) The extensive use of the calculus (differentiation and integration) and the use of complex numbers or solving differential equations should not be required to solve the theoretical and practical problems.

b) Questions may contain concepts and phenomena not contained in the Syllabus but sufficient information must be given in the questions so that candidates without previous knowledge of these topics would not be at a disadvantage.

c) Sophisticated practical equipment likely to be unfamiliar to the candidates should not dominate a problem. If such devices are used then careful instructions must be given to the candidates.

d) The candidates should know the system of units used in the country of origin (the original texts of the problems have to be set in the SI units).

e) The candidates should be familiar with the material

covered by the past problems of the International Physics
Olympiads.

A. Theoretical Part

Adopted in Portorož, Yugoslavia, June 1985

*The first column contains the main entries while the second
column contains comments and remarks if necessary.*

1. Mechanics

a) Foundation of kinematics of
   of a point mass

   | Vector description of the
   position of the the point
   mass, velocity and
   acceleration as vectors

b) Newton's laws, inertial
   systems

   | Problems may be set on
   changing mass

c) Closed and open systems,
   momentum and energy, work,
   power

d) Conservation of energy,
   conservation of linear
   momentum, impulse

e) Elastic forces, frictional
   forces, the law of
   gravitation, potential
   energy and work in a
   gravitational field

   | Hooke's law, coefficient of
   friction (F/R = const),
   frictional forces static and
   kinetic, choice of zero of
   potential energy

f) Centripetal acceleration,
   Kepler's laws

2. Mechanics of Rigid Bodies

a) Statics, center of mass, | Couples, conditions of
   torque | equilibrium of bodies

b) Motion of rigid bodies, | Conservation of angular
   translation, rotation, | momentum about fixed axis
   angular velocity, angular | only
   acceleration, conservation
   of angular momentum

c) External and internal | Parallel axes theorem
   forces, equation of motion | (Steiner's theorem)
   a rigid body around the | additivity of of the moment
   fixed axis, moment cf | of inertia
   inertia, kinetic energy of
   a rotating body

d) Accelerated reference | Knowledge of the Coriolis
   systems, inertial forces | force formula is not
                                | required

3. Hydromechanics

No specific questions will be set on this but students would be expected to know the elementary concepts of pressure, buoyancy and the continuity law.

4. Thermodynamics and Molecular Physics

a) Internal energy, work and | Thermal equilibrium,
   heat, first and second laws | quantities depending on
   of thermodynamics | state and quantities

40

| depending on process

b) Model of a perfect gas, | Also molecular approach to
pressure and molecular | such simple phenomena in
kinetic energy, Avogadro's | liquids and solids as boiling,
number, equation of state | melting etc.
of a perfect gas, absolute
temperature

c) Work done by an expanding | Proof of the equation of the
gas limited to isothermal | adiabatic process is not
and adiabatic processes | required

d) The Carnot cycle, | Entropy as a path independent
thermodynamic efficiency, | function, entropy changes and
reversible and irreversible | reversibility, quasistatic
processes, entropy | processes
(statistical approach),
Boltzmann factor

5. Oscillations and waves

a) Harmonic oscillations, | Solution of the equation for
equation of harmonic | harmonic motion, attenuation
oscillation | and resonance — qualitatively

b) Harmonic waves, propagation | Displacement in a progressive
of waves, transverse and | wave and understanding of
longitudinal waves, linear | graphical representation of
polarization, the classical | the wave, measurements of
Doppler effect, sound waves | velocity of sound and light,
| Doppler effect in one
| dimension only, propagation
| of waves in homogeneous and
| isotropic media, reflection
| and refraction, Fermats

|principle

c) Superposition of harmonic waves, coherent waves, interference, beats, standing waves

Realization that intensity of wave is proportional to the square of its amplitude. Fourrier analysis is not required but candidates should have some understanding that complex waves can be made from addition of simple sinusoidal waves of different frequencies. Interference due to thin films and other simple systems (final formulae are not required), superposition of waves from secondary sources (diffraction)

6. Electric Charge and Electric Field

a) Conservation of charge, Coulomb's law

b) Electric field, potential, Gauss' law

Gauss' low confined to simple symmetric systems like sphere. cylinder. plate etc.. electric dipole moment

c) Capacitors, capacitance, dielectric constant, energy density of electric field

7. Current and Magnetic Field

a) Current, resistance, internal resistance of

Simple cases of circuits containing non-ohmic devices

source, Ohm's law, Kirchhoff's laws, work and power of direct and alternating currents, Joule's law

with known V–I characteristics

b) Magnetic field (B) of a current, current in a magnetic field, Lorentz force

Particles in a magnetic field, simple applications like cyclotron, magnetic dipole moment

c) Ampere's law

Magnetic field of simple symmetric systems like straight wire, circular loop and long solenoid

d) Law of electromagnetic induction, magnetic flux, Lenz's law, self-induction, inductance, permeability, energy density of magnetic field

e) Alternating current, resistors, inductors and capacitors in AC–circuits, voltage and current (parallel and series) resonances

Simple AC–circuits, time constants, final formulae for parameters of concrete resonance circuits are not required

8. Electromagnetic waves

a) Oscillatory circuit, frequency of oscillations, generation by feedback and resonance

b) Wave optics, diffraction
from one and two slits.
diffraction grating,
resolving power of a
grating, Bragg reflection

c) Dispersion and diffraction
spectra, line spectra of
gases

d) Electromagnetic waves as | Superposition of polarized
transverse waves, | waves
polarization by reflection,
polarizers

e) Resolving power of imaging
systems

f) Black body, | Planck's formula is not
Stefan-Boltzmann's law | required

9. Quantum Physics

a) Photoelectric effect, | Einstein's formula is
energy and impulse of the | required
photon

b) De Broglie wavelength.
Heisenberg's uncertainty
principle

10. Relativity

a) Principle of relativity;
addition of velocities,

relativistic Doppler effect|

b) Relativistic equation of
   motion, momentum, energy,
   relation between energy and
   mass, conservation of
   energy and momentum

11. Matter

a) Simple applications of the
   Bragg equation

b) Energy levels of atoms and
   molecules (qualitatively),
   emission, absorption,
   spectrum of hydrogenlike
   atoms

c) Energy levels of nuclei
   (qualitatively), alpha-,
   beta- and gamma-decays,
   absorption of radiation,
   halflife and exponential
   decay, components of
   nuclei, mass defect,
   nuclear reactions

B. Practical Part

(Adopted in London-Harrow, United Kingdom, July 1986)

The Theoretical Part of the Syllabus provides the basis
for all the experimental problems. The experimental problems

given in the experimental contest should contain
measurements.

Additional requirements:

1. Candidates must be aware that instruments affect
   measurements.

2. Knowledge of the most common experimental techniques for
   measuring physical quantities mentioned in Part A.

3. Knowledge of commonly used simple laboratory instruments
   and devices such as calipers. thermometers, simple volt-.
   ohm- and ammeters, potentiometers. diodes, transistors,
   simple optical devices and so on.

4. Ability to use, with the help of proper instruction, some
   sophisticated instruments and devices such as double-beam
   oscilloscope, counter, ratemeter. signal and function
   generators, analog-to-digital converter connected to a
   computer, amplifier, integrator, differentiator, power
   supply, universal (analog and digital) volt-. ohm- and
   ammeters.

5. Proper identification of error sources and estimation of
   their influence on the final result(s).

6. Absolute and relative errors. accuracy of measuring
   instruments, error of a single measurement. error of a
   series of measurements. error of a quantity given as a
   function of measured quantities.

7. Transformation of a dependence to the linear form by
   appropriate choice of variables and fitting a straight
   line to experimental points.

8. Proper use of the graph paper with different scales (for

example polar and logarithmic papers).

9. Correct rounding off and expressing the final result(s) and error(s) with correct number of significant digits.

10. Standard knowledge of safety in laboratory work. (Nevertheless, if the experimental set-up contains any safety hazards the appropriate warnings should be included into the text of the problem.)

ANALYSIS

O N D E G R E E O F D I F F I C U L T Y O F T H E O L Y M P I C P R O B L E M S

The article by Gau and Gau [1] presented below contains some considerations on how to measure quality of the physics problems, in particular the olympic problems.

It is proper to mention here that the first attempt of this kind was made by G. S. Tarasyuk from the Moscow State University in her famous lecture "Исследование международных олимпиад по физике как средства развития творческих способностей учащихся" [2] given in Varna during the XII IPhO. The Russian text of the lecture was distributed among the delegation leaders. Unfortunately, at present it is practically unavailable. The approach by Tarasyuk was entirely different than that by Gau and Gau. Tarasyuk introduced the, so called, coefficient of difficulty determined by the results reached by the pupils solving the problems. The coefficient of difficulty of a given problem is defined as the ratio of 60% of the maximum number of points theoretically available for solving the problem to the mean number of points gained by all the participants for this problem. The coefficient of difficulty K of a given Olympiad is defined in a similar way. In her lecture Tarasyuk presented changes of the coefficient K of subsequent Olympiads as well as changes of this coefficient in time for the problems concerning different parts of physics separately (e.g. separately for the problems on optics, separately for the problems on electricity, etc.). For obvious reasons Tarasyuk analyzed only the first eleven Olympiads. The coefficients K of difficulty of these Olympiads are reproduced in the Fig. 1.

The coefficient K characterizes a complex of three items: problems + population of participants + marking teams. It is a very important coefficient. However, it gives us information or adequacy or non-adequacy of the problems after the competition, i.e. too late from the point of view

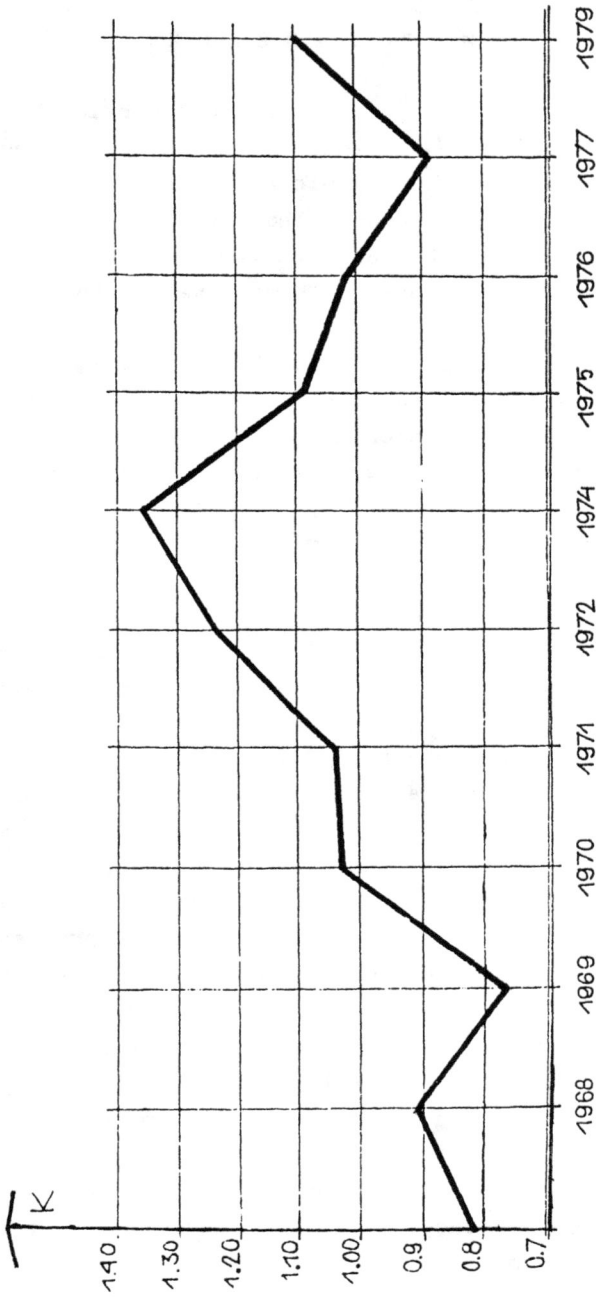

Fig. 1. The coefficients of difficulty K of the first eleven International Physics Olympiads [2]

of the organizers. On the other hand, Gau and Gau try to characterize the problems alone by assigning them a number of formal characteristics described in the article.

Which way is better for characterization of the olympic problems, in particular, for characterization of their adequacy to the olympic purposes? Probably neither is entirely satisfying as they do not consider, or do so only partly, many aspects of the olympic problems such as for example:

1. beauty of the problem;
2. is the problem original?
3. is the problem interesting?
4. is the problem instructive?
5. length of the text of the problem;
6. possibility of solving within framework of the Syllabus
7. is the problem well formulated?
8. creative possibility required for solving;
9. is the way of solving indicated in the text?
10. possibility of solving in different ways;
11. does the problem allow to establish sharply outlined difference between better and worse participants?
12. spectrum of knowledge tested by all the problems at a given competition;

etc., etc.

For the above mentioned examples of different aspects of the competition problems we think that the quality of the problems cannot be expressed with a single number. We are sure that a many-dimensional (multi-parameter) measure would be much better.

Good competition problem can be compared to a verse in poetry or to a painting in pictorial art. Everybody knows well that value of these items cannot be measured with a single number. Nevertheless, paintings have some, almost commonly accepted, price important for practical reasons.

So, although we realize that a single parameter
characteristics of the problem cannot be satisfactory,
certainly we shall look for some one-parameter "value" of
the competition problems that would be commonly accepted for
practical purposes.

We think that the article by Gau and Gau is very
interesting and we do hope that it will stimulate farther,
fruitful discussion on the olympic problems and on their
characterization. We invite you to join the discussion. The
best way is to send your analytic article directly to the
Editor. We are going to include some of them (or all,
depending on our possibilities) in the next volume.

At the end we would like to suggest that you compare
the coefficients introduced by Tarasyuk and by Gau and Gau.
Such a comparison will certainly be interesting for you.

The Editor
Waldemar Gorzkowski
Instytut Fizyki PAN
al. Lotników 32/46
02-668 Warszawa
POLAND

[1] B. Gau, R. Gau, next article in this issue.
[2] G. S. Tarasyuk, Исследование международных олимпиад по
физике как средства развития творческих способностей
учащихся, manuscipt of the lecture given in Varna during
the XII IPhO (distributed among the delegation leaders).

ON ALTERATIONS IN THE STRUCTURE AND REQUIREMENT LEVEL
OF THEORETICAL PROBLEMS SET
IN INTERNATIONAL PHYSICS OLYMPIADS

Barbara Gau and Rudolf Gau

Pädagogische Hochschule "Liselotte Herrmann" Güstrow

Sektion Mathematik/Physik

Goldberger Str. 12

2600 - Güstrow / GDR

1. INTRODUCTION

Currently, in many countries of the world great efforts are made in order to recognize talented pupils as soon as possible and to promote them systematically. This has its roots in humanistic concern to develop such pupils' individuality, to provide them routes and facilities of exhausting their optimal performances and of matching their wits with each other in competitions.

These exertions find their expression in permanently increasing number of participants in international olympiads for pupils, including the International Physics Olympiad (IPhO)[2].

Analysing the development of the IPhO we can realize that in addition to the increasing number of participants a further development in the content level of the problems can be stated. These alterations in the content level are characterized by:

- an enlargement with respect to the branches of physics the problembs are taken from (classical physics, quantum physics, nuclear physics, solid state physics, special relativity theory etc.),

- an increasing requirement level of the problems set (i.e. alterations in the level and in the structure of the problems) which is characterized by:

* a quantitative increase of the characteristics determining the requirement level of the problems set,
* an increase of creative elements in solving the
  problems
* an enlargement of mathematical requirements, and
* the application of complicated technology for solving
  the experimental problems (e.g. laser devices, oscillos-
  copes etc.).

Alltogether, both the physical and the mathematical complexity of the problems have increased in course of the
development of the IPhO.

Hence it has become obvious that a successful participation
in international physics competitions can only be secured by
special long-term preparation, since the requirements oc-
curring in the problems exceed the educational objectives of
physics instruction in the secondary schools of many countries.

2. ON THE PREPARATION OF PARTICIPANTS FOR THE INTERNATIONAL
   PHYSICS OLYMPIAD

Top results have been obtainted more and more by those
pupils having available a wide knowledge of all fields of
physics and, furthermore, showing extraordinary abilities
and aptitudes for solving theoretical and experimental
problems.

Consequently, particular emphasis is laid on

- the development of a high intellectual mobility of the
  pupils,
- the pupils' equipment with a broad, deep and in many
  cases, special knowledge,
- the pupils' instruction in subject-specific ways of
  thinking and working.

This can only be reached within a long-term process which is
uniformly and continueosly led. It covers the provision of

an extensive knowledge of special physical subjects on a high level in close connection with solving problems and discussing solution variants.

Thus, the conception for promoting talented pupils in their preparation for international physics olympiads has received a qualitative development. In the first years of the olympiads emphasis in the pupils' preparation was laid on solving many demanding problems (in the sense of turning from quantity to the next higherest quality), whereas in recent years the significance of providing systematic knowledge on a high level increased more and more.

International investigations prove the significant effect of knowledge in order to solve intellectually pretentious problems in all scientific fields. In 1986 VAN DER MEER[5] stressed that a fully developed availability of operative factors is characteristic of gifted pupils. "The acquisition of qualitatively excellent knowledge becomes determining for the advance in the achievment level of highly talented."

In 1985 WEINERT/WALDMANN[6] explained that the change from a beginner to a specialist, however, cannot be considered as permanent development of knowledge. "Beginners have simple, stable, broadly applicable information processing strategies, which, however, will not work, if the problems are complex and unifamiliar to the pupils. With growing experience the perceptual and cognitive schemes become more complex."

That means, the prerequisite for a successful development of talents is excellent knowledge in connection with experience in solving problems with high requirements. Similar observation could be made in our investigations during the preparation of gifted pupils for international physics competitions, too.[1]

In order to qualify this process of developing highly gifted pupils in the field of physics, beside the question of organization and contents that question is of interest how we succeed in setting always high and contineuosly growing demands on the pupils which promote their development.

In our investigations we concentrate on requirements the pupils have to meet in the process of solving physical problems. Physical problems activate and stimulate the pupils for increasingly creative thinking. This can -for instance- be reached by such problems in which the finding of the solution and the construction of the physical model is not possible by means of algorithms.
In their totality physical problems have a multi-potentiality in the development of pupils gifted in physics. They support the planned and systematic acquisition and implant of physical knowledge and abilities of the pupils. Especially the application on problems relevant to practice promotes not only the understanding of physical facts but contributes also to make conscious the necessity and possibility of using the acquired knowledge to solve these problems.
Solving physical problems promotes also the education of the pupils while persueing in aim and critical evolution of their own work, perseverance, exactness and cleanliness are required. Moreover, solving of problems is important for the development of a scientific ideology of life (universal validity of natural laws, unit of the material world, enthusiasm for scientific-technological progress, etc.).
For the qualification of the pupils in order to solve demanding problems one must known characterize the requirements. Only if objectivly can predicted what to what degree is demanded from a pupil one has the control of the "key", how to increase the requirements contineuosly.
I.e. one has to look for the elements and characteristics respectively which characterize the requirements of a problem and allow a comparison between different problems. In

this way, problems of compareable level can be set in the preparation phase of IPhO's. Simultaneously an apparatus is available for describing mor exactly what elements have caused an increase of requirement level in the course of the development of the IPhO. In the same way trends of this development can be red off.

After the XV. IPhO 1984 in Sigtuna (Sweden) all leaders of the delegations agree, that all problems in this olympiade not only were more difficult as in earlier olympiads, but also qualitatively "new" requirements contained. This new requirement are characterized by the creativity which the solutions demaned on the pupils.

For instance an example: we want to discuss the second problem of these olympiade, where the pupils must construct independently a physical model for solving this problem. We will use this problem to explain our model to analyse the structure of physical problems and to find the characteristics which determine the requirements.

Problem 2. (XV. IPhO 1984) [4] *:*

In some lakes a curious phenomenon "seiching" (oscillation of water) can be observed occasionally. It is usally seen an shallow, long and comparatively narrow lakes. The whole mass of the water seems to move like the coffee in a cup when you carry it to a waiting guest, which can not be mistaken for the waves normally seen on lakes.

To create a model we take a rectangular vessel. Denote the length of the vessel by L and the height of the water by h. Assume that the water level initially includes a small angle with the horizontal. The water level beginns to oscillate in such a way that it remains always planar but oscillates about a horizontal axis half way along the length of the vessel.

a) Develop a model for the motion of the water and find an expression for the period of oscillation T. The initial conditions are given in the fig.1. Assume that ξ << h.

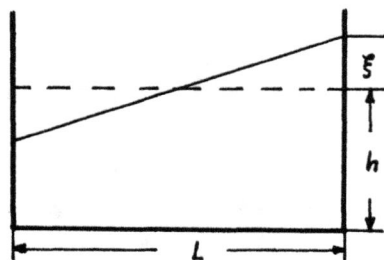

Fig.1

L = 479 mm		L = 143 mm	
h (mm)	T (s)	h (mm)	T (s)
30	1.78	31	0.52
50	1.40	38	0.48
69	1.18	58	0.43
88	1.08	67	0.35
107	1.00	124	0.28
124	0.91		
142	0.82		

b) Our tables show oscillation-times for different water depths in two vessels of different lengths. Check in a suitable way how well your formula fits the experimental data and give your opinion on the usefulness of the model.

c) Fig.2 shows some results from measurements on Lake Vättern in Sweden at Bastedalen (northern end) and Jönköping (southern end). The lake has a length of 123 km and an average depth of 50 m. What is the time scale in this case?

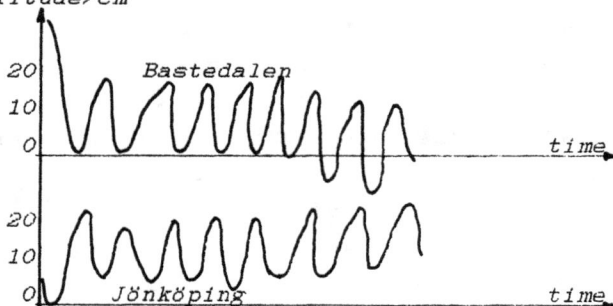

Fig.2

3. THE MODELLING OF THE DETERMINATION AND FORMATION OF REQUIREMENTS

Requirements comprise the totality of objective requirements of the activity of pupils when solving problems. According to HACKER 1976 they have to be considered as a function of the problem, the given realization conditions and the necessary and real existing efficiencies of pupils (fig.3).

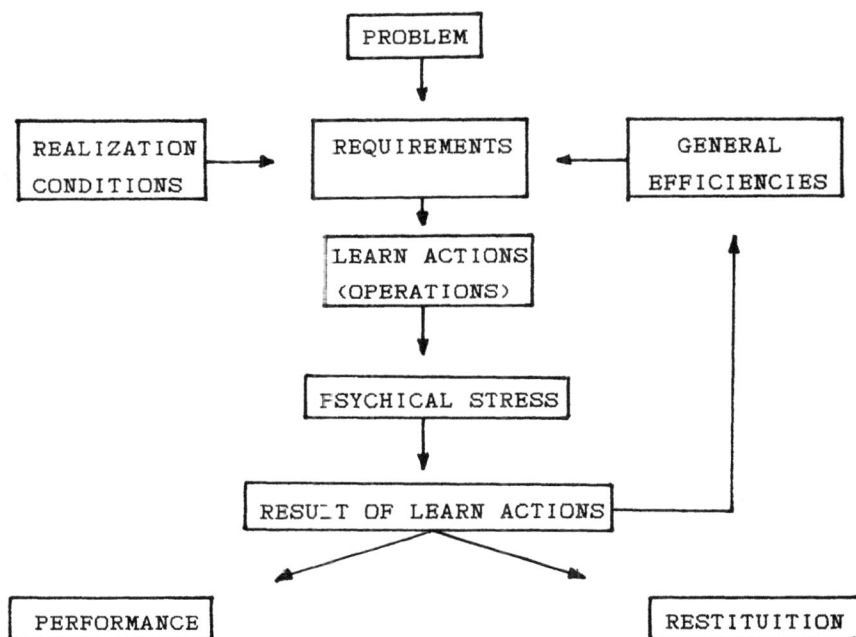

```
                        ┌──────────────┐
                        │   PROBLEM    │
                        └──────┬───────┘
                               │
                               ▼
┌──────────────┐        ┌──────────────┐        ┌──────────────┐
│ REALIZATION  │───────▶│ REQUIREMENTS │◀───────│   GENERAL    │
│ CONDITIONS   │        └──────┬───────┘        │ EFFICIENCIES │
└──────────────┘               │                └──────────────┘
                               ▼                        ▲
                        ┌──────────────┐                │
                        │ LEARN ACTIONS│                │
                        │ (OPERATIONS) │                │
                        └──────┬───────┘                │
                               │                        │
                               ▼                        │
                        ┌──────────────┐                │
                        │PSYCHICAL STRESS│              │
                        └──────┬───────┘                │
                               │                        │
                               ▼                        │
                  ┌────────────────────────┐            │
                  │ RESULT OF LEARN ACTIONS │───────────┘
                  └───────┬─────────┬──────┘
                         ╱           ╲
                        ▼             ▼
        ┌──────────────┐        ┌──────────────┐
        │ PERFORMANCE  │        │ RESTITUITION │
        └──────────────┘        └──────────────┘
```

Fig.3:

The quantitative impression is given by the requirement level of a problem. One has to differ between the level and the complication degree of a problem.

The complication degree is a subjective category. It characterizes the relation between the objective requirements and the real existing efficiencies of the single pupil. The complication degree of a problem indicates the probability by which the pupil <u>cannot</u> solve the problem correctly [1].

A determination of the totality of the objective activity requirements of a pupil when solving a problem is only possible using a suitable idealization, that means modelling, because of the complexity of the problem set.

Our model for the determination of requirement level is based on the following assumptions:

1. The general process of solving physical problems is considered beeing a sequence of linear connected operations.

2. It is not possible to resolve the solving process into a unique sequence of operations. Therfore indicators have to be used.
   As suitable indicators we consider problem characteristics, in the following called <u>requirement determing characteristics.</u>

3. The analysis of requirements is reduced to the analysis of the characteristics of a problem. It is abstracted from the real efficiency assumptions of the pupil (his actual knowledge, way of thinking, motives, habits etc.). Just as before we do not consider the influence of different realization conditions on the requirements. With that the investigations are based on a simplified function model of the relations between requirements, demand and result of the learning activity (fig.3).

4. Requirements can be considered as a system. They have a structure that consists of a set of elements (requirement determing characteristics) and their mutual relations. It must be possible to determine these characteristics from the solution and to differ between them.

5. The deepening of some characteristics can only be determined qualitatively. To this belong:
   * the cognition plane in which the facts are given by

words and figuretively (percibtibly-concrete, figura-
tively, figuratively-abstract) and

* the degree of abstraction of the law statements and
the problem objective and the differences between both
of them.

(These characteristics we don't consider no more.)
However the majority of requirement determing characteristics can be given quantitatively, that means determined
numerably from a solution.
The characterization of the requirement determing characteristics shall be done using the solution of the above
mentioned problem of the XV. IPhO.

Solution of problem 2 (XV. IPhO 1984) [4] :
*a) Denote the horizontal and the perpendicular displacement
of the centre of gravity of the water by x and y respectively. The mass of water confined in equilibrium by
triangle PCA moves in the extreme phase of oscillation into
the space confined by triangle PBD (see fig.4). Its centre
of gravity is displaced by 2L/3 in horizontal and by $2\xi/3$
in vertical direction. the ratio of the displaced water to
the total mass is given by*

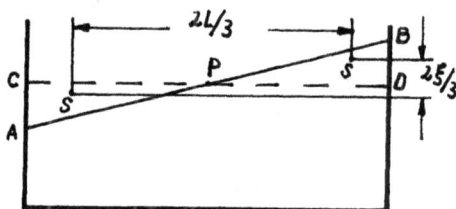

Fig. 4:

*The displacement of the centre of gravity of the total mass
of water relative to its position in equilibrium is*

$$\Delta x = \frac{\xi}{4h} \cdot \frac{2L}{3} = \frac{\xi \cdot L}{6h}$$ *in horizontal* (1)

and

62

$$\Delta y = \frac{\xi}{4h} \cdot \frac{2\xi}{3} = \frac{\xi^2}{6h} \qquad\qquad in\ vertical \qquad (2)$$

direction. The velocity of the centre of mass then follows by derivation:

$$V_x = \frac{\dot{\xi}L}{6h} \quad , \quad V_y = \frac{2\xi\dot{\xi}}{6h} \qquad\qquad thus \quad V_x \gg V_y$$

The motion of water is therefore primarily horizontal. We work out a model neglecting the motion of water relative to, the centre of mass.

We assume that the horizontal motion of water is a harmonic oscillation and the whole mass of water moves together. Denoting the maximum velocity of the horizontal motion by v and the angular frequency of the oscillation by ω, we have

$$V = \Delta x \cdot \omega \qquad\qquad (3)$$

Let us suppose further that during oscillation the whole mass of water moves horizontally, when the surface moves horizontally. This is a reasonable assumption because h is much smaller than L. Thus

$$\frac{1}{2} m \cdot v^2 = m \cdot g \cdot \Delta y$$

where m is the whole mass of water. From equations (1), (2), (3) we get for the period of oscillation:

$$T = \frac{\pi \cdot L}{\sqrt{3 \cdot h \cdot g}} \qquad\qquad (4)$$

b) The numerical results using (4) for given heights of water we can compare with the experimental data. Plotting T against h the points are on a straight line. From the diagramm it is easily seen that the numerical data agree well measured data but there is a systematic error of oca 15 %. It is therefore advisable to use a formula with a corresponding correction:

$$T = \frac{1,15 \cdot \pi \cdot L}{\sqrt{3 \cdot h \cdot g}} ,$$

$$(5)$$

c) Using (5) the error will be less than 10 %. For the period of oscillation in the case of Lake Vättern we get T = 3.2 hours.

Many other models giving roughly the same results are possible and will of course also be accepted.

Our investigations of the requirement level of the IPhO-problems proved the following requirement determing characteristics being essential [1].

1. Branches of physics from which facts have been included:
In the example only facts from mechanics are required while the trend in the development shows that facts of several branches often are combined in a problem.

(number of characteristics: 1)

2. Transformations of common language concepts into those of subject language that will be necessary to describe the physical model of the problem:
From the demand "Develop a model for the motion of the water...", of part a) of the problem results the question what is a "model" for the motion. In part b) of the problem is demanded, "check in a suitable way...". One has to ask what does this mean. The pupils must recognize how to compare the facts given in the problem with their own results (calculation of deviations and suitable diagrams respectively).
Furthermore one has to recognize by what the efficiency of the model can be tested. In case of an example it is possible by the discussion of the deviations between the theoretical and experimental results. A suitable representation has to be found independently. Another demand consists of the recognition of the necessity to determine the required scale by means of the own model (part c).

(number of characteristics: 4)

3. Special terms to be used that arn't given in the text but that are absolutely necessary for the solution:
The oscillation of water is interpreted as a harmonic oscillation of the centre of gravity. That means the concept centre of gravity is absolutely necessary for the modelling. Another model could be, for example, that the water oscillates like a rigid solid around a fixed axis of rotation.

(number of characteristics: 1)

4. Conditions to be assumed necessarily:
Because of the small deviation ($\xi <<$ h) the loss due to the friction can be neglected and the motion can assumed to be a harmonic one. Only this assumption enables the pupil to find an analytic solution.

(number of characteristics: 1)

5. Conditions to be assumed necessarily that lead to simplifications:
In the given solution is assumed that the alteration of the centre of gravity in vertical direction is much more smaller than the alteration in horizontal direction (x >> y). From this reults a simplification for the description of the motion of the centre of gravity.
Such conditions are, the approximation sin x = x for smal angles or the approximation of a function by a Taylor series.

(number of characteristics: 1)

6. Conditions to be assumed necessarily that lead to distinguish between different cases:
This requirement does not appear in the given example.
A typical example for such requirements is problem 3 of the same IPhO, in which different possibilities to link construction elements have to be investigated [4), 7)].

7. Number of laws that must be included to find the solution:
Here the laws of the single branches of physics are counted that correspond to the actual level. Definition equations of physical quantity and laws from the beginner-lessons in physics are excluded.
In the given problem we have
- law for the harmonic oscillation,
- law for the calculation of the centre of gravity.
(By using of another model the number of characteristics can be higher, e.g. calculation of the moment of inertia, law of Steiner).

(number of characteristics: 2)

8. Principles and laws of conservation that are necessary for the solution:
In the considered solution it is necessary to apply the energy conservation law for determination of the oscillation-time of the water.

(number of characteristics: 1)

9. Number of intermediate quantities to be calculated:
To get a complete solution in the given problem 8 quantities have to be calculated. This number must be considered a minimum because by use of another model the number of this characteristics can rather increase.

(number of characteristics: 8)

10. Expense of numerical calculations:
The degree of this requirement on the level of IPhO-problems can only be estimated. In our model we used the following classification:
(1) ==> in case of algebrical calculations;
(2) ==> in case of trigonometrical, logarithmic calculations and diagrams;

(3) ==> in case of calculations involving complex numbers, grafical solutions etc..

(degree of requirements: 2)

11. Applications of mathematical relations:

Here we classify,

(0) ==> in case of transforming equations,

(1) ==> solving linear and quadratic equations and systems of equations with two variables,

(2) ==> Solving equations of 3th. degree and systems of equations of more as two variables etc.,

(3) ==> Application of addition theorems, differential and integral calculus, sequences and series etc..

(degree of requirement: 3)

By means of this model for the determination of the requirements of theoretical problems of physics competitions all problems of the IPhO's have been investigated in order to draw conclusions for a systematic and effective preparation of participants for international physics competitions.

4. RESULTS AND CONCLUSIONS

Knowledge gained by means of a model posses a hypothetical character. With this the theoretically determined requirements of olympiad problems have to be related to the real mastering of the problems by the participants. Only if one can prove the existing of a relation between the theoretical and empirical findings the requirement determing characteristics can be used in their totality in order to predict the complication degree of the problem for further participants of competitions.

To test this relation it is necessary to develop a mathematical model of the problem. Already in case of the simplest model of the requirement level as sum of the frequencies X_i of all requirement determing characteristics i

$$A = \sum X_i$$

a good correlation between the requirement level and the complication degree can be stated. Because of the simplicity of the model the conclusions to be drawn from the model are limited. Therefore the discussion of the requirement level must always be linked up with a clarification of the structure of the problem.

The number of requirement determing characteristics, the requirement level for each problem and the average for the theoretical problems of the XV. IPhO 1984 are indicated in the following table.

	Nr.1	Nr.2	Nr.3
1. Branches of physics from which facts have included	2	1	1
2. Transformations of common language concepts into those of subject language	3	4	4
3. Special terms to be used that arn't given in the text	0	1	0
4. Conditions to be assumed necessarily	1	1	2
5. Conditions to be assumed necessarily that lead to simplifications	1	1	2
6. Conditions to be assumed necessarily that lead to distinguish between cases	0	0	4
7. Number of laws that must included to find the solution	3	2	2
8. Principles and laws of conservation that are necessary for the solution	1	1	0
9. Number of intermediate quantities to be calculated	2	8	5
10. Expense of numerical calculations	1	2	1
11. Applications of mathematical relations	2	3	3
	A=15	A=24	A=24

The average requirement level: $\bar{A} = 21$

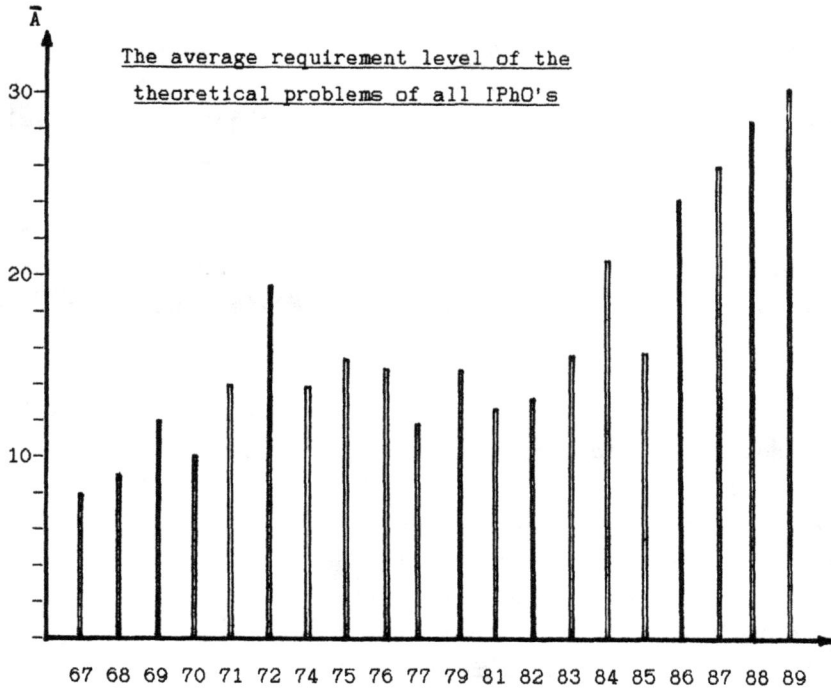

Fig.5:

Particular results are the following:

1. From the diagram (fig. 5) representing the average requirement level of the theoretical problems alterations of the level can be stated that permit a classification in the following stages:

* 1967 till 1970: a gradually increasing,
* 1971 till 1983: $\bar{A}$ remains nearly at the same level,
* 1984 till 1989: a permanent increasing.

2. The question by what the increase of the level had been caused leads to investigate the contribution of single requirement determing caracteristics to the increase [1].

From 1971 until 1983 compared with the initial years of the IPhO's the major increase is given by the application of conservation laws. The graduation is continued with conditions that lead to different cases an to intermediate calculations.

With that it is a question of characteristics that have been called typical for the development of creative activities. Emphasize is to put on the constancy of the numerical calculation and mathematical relations on the lowest stage to be applied.

Since 1984 the major increase in comparison with the former period results from conditions to be assumed and transformation of common language terms into special subject terms and necessary conceptions respectively which do not appear explicitly in the text of the problem. I.e., the creative components grown constantly in the development of the requirements. The increase of the requirement resulting from the mathematical complexity reaches the same dimension. This is an essential quantitative and qualitative alterations in comparison with the former period.

3. For the pupil's preparation for the IPhO's a balanced relation between the acquisition of knowledge and training is necessary. In the theoretical lessons the structure and the laws have to be represented; methods of the physical description must be explained. This provides all pupils with an integrated theoretical knowledge that corresponds to the requirements.

Their knowledge is systematized and completed and ideas for a sytematic self instruction are given. In the training the acquired knowledge is consolidated and deepened. Using concrete problem solution strategies are explained and trained.

Firm knowledge of laws, including the laws of conservation, their range of validity, experiences in modelling physical events and processes and formed abilities while working mathematically on problems have to be developed as a fundamental assumption for succesful solution of problems of high standard.

4. With the characterization of the structure of requirements of problems set in IPhO's it is possible to test existing problems (i.e. the problems of the national competitions) for serving for the preparation of the pupils with respect to the competitions, to construct problems or varying given problems.
In this way the preparation process of the pupils can be improved.
This holds principally for all promoton forms of highly-gifted pupils independent of the promotion for national or international competitions.

5. The analyse of all former theoretical problems of the International Physics Olympiads shows clearly the accentuation move of the requirement structure of the problems and thus permits a prediction of possible requirements of future olympiads.

So the model for the determination of the characteristics determing the requirements of physical problems can justify its methodical function in the process of compiling problems and exercises and trend statements of the development level of the theoretical problems of IPhO's.

REFERENCES

[1] Gau, B., "Erkennen und Fördern begabter Schüler auf dem Gebiet der Physik unter Berücksichtigung der Gestaltung von Anforderungen physikalischer Aufgaben", Diss. B., Güstrow (1988).

[2] Gorzkowski, W., "International Physics Olympiads - History and Perspectives", (published in this volume).

[3] Hacker, W., "Psychische Regulation von Arbeitstätigkeiten: Innere Modelle, Strategien in Mensch-Maschine-Systemen", Psychische Regulation von Arbeitstätigkeiten, 21-38 (1976).

[4] Kunfalvi, R., "Collection of Competition Tasks from the 1st trough XVth International Physics Olympiads 1967-1984", Budapest (1985).

[5] Van der Meer, E., "Experimentelle Begabungsforschung", Dt. Verlag der Wiss., Berlin 67-79 (1986).

[6] Weinert, F.E. and Waldmann, M.R., "How do the gifted think: Intellectual abilities and cognitive processes", 6. Weltkongreß über begabte und talentierte Kinder, Hamburg (1985).

[7] Wendt, J., Gau, R., Walta, U., "Aufgabensammlung Physik - Physikolympiaden", Berlin (1987).

WINNERS

L I S T   O F   W I N N E R S

I International Physics Olympiad
Warszawa (Poland) 1967

Prizes:

1. Sandor Szalay	Hungary
2. Jaroslav Kozacik	Czechoslovakia
3. Bogdan Cichocki	Poland
4. Laszlo Mihaly	Hungary
5/7. Georgi Georgiyev	Bulgaria
5/7. František Klein	Czechoslovakia
5/7. Roman Libuszewski	Poland

Honourable Mentions:

1. Ferenc Marossy	Hungary
2. Władysław Skarbek	Poland
3. Ladislav Obdrżalek	Czechoslovakia
4. Aurelian Predeasa	Romania

Best total score:

Sandor Szalay	Hungary

II International Physics Olympiad
Budapest (Hungary) 1968

I Prizes:

1. Tomasz Kręglewski	Poland
2. Mojmir Simersky	Czechoslovakia

II Prize:

1. Laszlo Takacs   Hungary

III Prizes:

1. Peter Georgiyev   Bulgaria
2. Emil Matev   Bulgaria
3. Laszlo Mihaly   Hungary

Honourable Mentions:

1. Andrej Detela   Yugoslavia
2. Ondrej Krivanek   Czechoslovakia

Best total score:

       Tomasz Kręglewski   Poland

T. Kręglewski and M. Simersky gained the same total number of points. T. Kręglewski, however, solved the experimental problem better than M. Simersky and the International Board decided to place his name first.

Special Prize:

* *for the best solution to the problems 2 and 4:*

       Joachim Loose   GDR

The above results were reconstructed due to kind help of: Ms. Galina Sergeyevna Tarasyuk (Soviet Union), Dr. Daniel Kluvanec (Czechoslovakia), Prof. R. Kunfalvi (Hungary), Prof. Joachim Wendt (GDR) and Nicola Velchev (Bulgaria).

III International Physics Olympiad
Brno (Czechoslovakia) 1969

I Prizes:

1.	Šob Mayer	Czechoslovakia
2.	Peter Karoly	Hungary
3.	Wolfgang Pils	GDR
4.	Vladimir Černy	Czechoslovakia
5.	Andrey Klimov	Soviet Union
6.	Antonije Džordžeić	Yugoslavia
7.	František Komin	Czechoslovakia
8.	Jozsef Schpitzer	Hungary
9.	Vladimir Gavrilenko	Soviet Union
10.	Petre Dan	Romania
11.	Peter Kalman	Hungary
12.	Marin Mitov	Bulgaria
13.	Aleksey Chernoutsan	Soviet Union

II Prizes:

1.	Klementz Müller	GDR
2.	Karel Šafažik	Czechoslovakia
3.	S. Sekachev	Bulgaria
4.	Vladimir Merkulov	Soviet Union
5.	Przemysław Prusinkiewicz	Poland
6.	Laszlo Auder	Hungary
7.	Juri Šafažik	Czechoslovakia
8.	Antal Jaricki	Yugoslavia
9.	Jan Oficjalski	Poland
10.	Peter Horvaty	Hungary

III Prizes:

1.	Joachim Loose	GDR
2.	Juliusz Poltz	Poland

3.	Zdravke Pribar	Yugoslavia
4.	Kraso Tisaj	Yugoslavia
5.	Joachim Bergmann	GDR
6.	Miroslav Dotečić	Yugoslavia
7.	Izak Bivas	Bulgaria
8.	Sorin Meskes	Romania
9.	Nikolay Kondratyev	Soviet Union

Honourable Mentions:

1.	Michael Josch	GDR
2.	Dimitr Danev	Bulgaria
3.	Svilen Bidikov	Bulgaria
4.	Feliks Przytycki	Poland
5.	Victor Birman	Romania

Best total score:

Šob Mayer   Czechoslovakia

Three first winners gained the same total number of points, but after analysis of the solutions the International Board decided that the solutions by Š. Mayer from Czechoslovakia were slightly better and placed his name in the first position.

The above results were reconstructed due to kind help of Ms. Galina Sergeyevna Tarasyuk from the Moscow University.

IV International Physics Olympiad
Moscow (Soviet Union) 1970

I Prizes:

1.	Mikhail Voloshin	Soviet Union
2.	Sergey Gorbachevskiy	Soviet Union

3. Borys Petrov   Soviet Union
4. Marek Ziółkowski   Poland

II Prizes:

1. Manfred Fischer   GDR
2. Antonie Dżordżević   Yugoslavia
3. Igor Luksyutov   Soviet Union
4. Andras Kerestury   Hungary
5. Jan Oficjalski   Poland
6. Igor Bulyzhenkov   Soviet Union
7. Pal Ormos   Hungary

III Prizes:

1. Miroslv Gandl   Czechoslovakia
2. Andras Nagy   Hungary
3. Vladimir Cherry   Soviet Union
4. Wojciech Ganzal   Czechoslovakia
5. Vladimir Kravcov   Soviet Union
6. Jaroslav Kučera   Czechoslovakia
7. Vaclav Novotny   Czechoslovakia
8. Dragosz Felie   Romania
9. Florin Curcudeu   Romania
10. Octavian Sima   Romania

Honourable Mentions:

1. Antoni Cieniek   Poland
2. Kornel Seiler   Hungary
3. Jon Apostol   Romania
4. Arkadiusz Rydel   Poland
5. Stanisław Chyrczakowski   Poland
6. Manfred Krzikalla   GDR
7. Władysław Majewski   Poland
8. Ivo Nachev   Bulgaria

9. Aleksander Zavalevski	Yugoslavia
10. Ognyan Nikolov	Bulgaria
11. Oliver Džurić	Yugoslavia
12. Branko Dolenc	Yugoslavia
13. Michael Josch	GDR

Best total score:

Mikhail Voloshin	Soviet Union

Special Prizes:

* for the best solution to the experimental problem:

Manfred Fischer	GDR

* for the only female participant:

Inge Reimann	GDR

The names of the winners were transcibed from the Russian alphabet.

V International Physics Olympiad
Sofia (Bulgaria) 1971

I Prizes:

1. Karel Šaverdžek	Czechoslovakia
2. Adale Tichy	Hungary
3. Ulf-Hendrick Gleser	GDR
4. Andrey Varlamov	Soviet Union
5. Ivan Gabas	Hungary

II Prizes:

1. Andrej Kugler   Czechoslovakia
2. Aleksey Abrikosov   Soviet Union
3. Tomas Moso   Hungary
4. Władysław Iwanowicz   Romania
5. Władysław Majewski   Poland
6. Jan Oficjalski   Poland

III Prizes:

1. Władysław Kisch   Romania
2. Teodor Dumitrescu   Romania
3. Marek Ziółkowski   Poland
4. Aleksander Rusiecki   Poland
5. Nikolay Ganev   Bulgaria
6. Ferenc Iglos   Hungary
7. Lajos Goocs   Hungary
8. Aleksander Snegiryev   Soviet Union
9. Klaus Vogler   GDR
10. Thomas Jung   GDR
11. Hans Schmidt   GDR
12. Viorel Jonescu   Romania

Honourable Mentions:

1. Teodor Mishonov   Bulgaria
2. Sergey Budnik   Soviet Union
3. Vaclav Holy   Czechoslovakia
4. Zoltan Sabo   Hungary
5. Johanes Kreuziger   GDR
6. Vitas Saldzhunas   Soviet Union
7. Todor Vulpan   Romania

82

Best total score:

        Karel Šaverdżek   Czechoslovakia

The names of the winners were transcibed from the Bulgarian alphabet.

VI International Physics Olympiad
Bucharest (Romania) 1972

I Prizes:

1. Zoltán Szabó   Hungary
2. Sergey Provotorov   Soviet Union

II Prizes:

1. Serge Bouc   France
2. Liviu Condriuc   Romania
3. Stefan Gh. Pentiuc   Romania

III Prizes:

1. Sergey Lyagushin   Soviet Union
2. Gyula Imhof   Hungary
3. Michel Poirier   France
4. Hans Jürgen Schmidt   GDR
5. Adrian Ocneanu   Romania
6. Andras Kover   Hungary
7. Ruben Mkrtchyan   Soviet Union
8. Tiberiu Tugui   Romania
9. Nándor Éber   Hungary
10. Andrzej Golnik   Poland
11. Roman Stępniewski   Poland
12. Jiri Binder   Czechoslovakia
13. Dieter Achilles   GDR

14. Aleksander Khristov Ganev   Bulgaria

Honourable Mentions:

 1. Mihail Marin   Romania
 2. Józef Dygas   Poland
 3. Roland Matthets   GDR
 4. Igor Pletnyev   Soviet Union
 5. Pavel Drabek   Czechoslovakia
 6. Frank Michael Missbach   GDR
 7. Lev Vaydman   Soviet Union
 8. François Bacchus   France
 9. Frank Fischer   GDR
10. Jiri Dolejši   Czechoslovakia
11. Janusz Miller   Poland
12. Lajos Gacs   Hungary
13. Milan Lehotsky   Czechoslovakia
14. Zdzisław Otwinowski   Poland
15. Libor Slezak   Czechoslovakia

Best total score:

      Zoltán Szabó   Hungary

Special prizes:

* *for the best solution to the experimental problem:*

      Hans Jürgen Schmidt   GDR

* *for the best solutions to the problem on mechanics:*

        Andrzej Golnik   Poland
        Józef Dygas   Poland
        Gyula Imhof   Hungary
        Zoltán Szabó   Hungary

* for the best solution to the problem on electricity:

Zoltán Szabó   Hungary

* for the best solution to the problem on optics:

Sergey Provotorov   Soviet Union

VII International Physics Olympiad
Warszawa (Poland) 1974

I Prizes:

1. Jarosław Deminet   Poland
2. Géza Meszéna   Hungary
3. Jerzy Tarasiuk   Poland

II Prizes:

not awarded

III Prizes:

1. Jörg Bergmann   GDR
2. Jiri Hruška   Czechoslovakia
3. Felix Kerstan   GDR
4. Anatoliy Kuchanov   Soviet Union
5. György Pálfalvi   Hungary
6. Aleksey Rudnyev   Soviet Union
7. Zdenek Svoboda   Czechoslovakia
8. Kàroly Vladár   Hungary

Honourable Mentions:

1. Ulf Brüster	GDR
2. Yevgeniy Falkin	Soviet Union
3. Hans—Georg Martin	GDR
4. Sergey Masich	Soviet Union
5. Oleg Naniy	Soviet Union
6. Mihai Ganciu-Petcu	Romania
7. Maria Titeica	Romania

Best total scores (ex aequo):

Jarosław Deminet	Poland
Jerzy Tarasiuk	Poland

Special prizes:

* *for the best solution to the problem on optics:*

Jerzy Tarasiuk	Poland

* *for the best score by a female participant:*

Maria Titeica	Romania

(The names of the winners in each category of prizes are ordered alphabetically)

VIII International Physics Olympiad
Güstrow (GDR) 1975

I Prizes:

1. Jörg Bergmann	GDR
2. Volker Fritzsche	GDR
3. Sergey Korshunov	Soviet Union

4. Miroslav Lyčka	Czechoslovakia
5. Radu Oprea	Romania
6. József Schmidt	Hungary
7. Matthias Wagner	GDR

II Prizes:

1. Leonid Avdeyev	Soviet Union
2. Dominique Delande	France
3. Béla Faragó	Hungary
4. Laurentiu Frangu	Romania
5. Martin Hanke	GDR
6. Jan Hula	Czechoslovakia
7. Antoni Łączkowski	Poland
8. Yevgeniy Shakhnovich	Soviet Union
9. Jenö Szép	Hungary

III Prizes:

1. Jerzy Bławzdziewicz	Poland
2. Vadim Boryu	Soviet Union
3. Pierre-Michel Delpeuch	France
4. René Godefrey	France
5. Géza Györgyi	Hungary
6. Jiri Hulka	Czechoslovakia
7. Andrzej Łusakowski	Poland
8. Yuriy Makedonov	Soviet Union
9. Hans-Georg Martin	GDR
10. Dariusz Pietraszkiewicz	Poland
11. Bogusław Sulikowski	Poland
12. Attila Wirosztek	Hungary

Honourable Mentions:

1. Frank Anton	FRG
2. Didier Mayou	France

3. Réne Moine   France
4. Valentin Popov   Bulgaria
5. Peter Simeonov   Bulgaria
6. Dymitr Stoyanov   Bulgaria
7. Mihai Tâzlâuanu   Romania
8. Jiri Vyskočil   Czechoslovakia

Best total score: [1]

     Sergey Korshunov   Soviet Union

Special prize:

* for the best solution of the problem on mechanics:

     Jerzy Bławzdziewicz   Poland

(The names of the winners in each category of prizes are ordered alphabetically)

IX International Physics Olympiad
Budapest (Hungary) 1976

I Prizes:

1. Vladimir Bulatov   Soviet Union
2. Ildar Khamitov   Soviet Union
3. Vladimir Krivtsun   Soviet Union
4. Krzysztof Kulpa   Poland
5. Rafał Łubis   Poland
6. Alain Poirson   France
7. Dragos Gheorge Popescu   Romania

----

[1]   *The information received thanks to kindness of Dr. Daniel Kluvanec from Czechoslovakia.*

II Prizes:

1. Alexandru Chiosea   Romania
2. Rolf Glaser   GDR
3. Andrey Golubenchev   Soviet Union
4. Mathias Hegner   GDR
5. Reinhard Meinel   GDR
6. Pavel Tarina   Czechoslovakia
7. Jiri Svoboda   Czechoslovakia

III Prizes:

1. Werner Becker   FRG
2. Mats Carlson   Sweden
3. Alexandru Dumitru   Romania
4. Lothar Köhler   FRG
5. Jean-Marc Luck-Laverne   France
6. Hans-Georg Martin   GDR
7. Svetlozar Nedev   Bulgaria
8. Cristian Panaiotu   Romania
9. Vanko Peter   Hungary
10. Kai-Uwe Posnecker   FRG
11. Sorina Saceanu   Romania
12. Ilja Turek   Czechoslovakia

Honourable Mentions:

1. Farago Bela   Hungary
2. Konrad Gajewski   Poland
3. Wojciech Jawień   Poland
4. Stilyan Kalichin   Bulgaria
5. Karel Kubat   Czechoslovakia
6. Gulyas Mihaly   Hungary
7. Thomas Müller   GDR
8. Minh Son Nguyen   France

9. Waldemar Rachowicz   Poland
10. Stoyan Rusev   Bulgaria
11. Valeriy Starshenko   Soviet Union
12. Jean-Marc Victor   France
13. Attila Virosztek   Hungary

Best total score (special prize):

    Rafał Łubis   Poland

(The names of the winners in each category of prizes are ordered alphabetically)

X International Physics Olympiad
Hradec Kralove (Czechoslovakia) 1977

I Prizes:

1. Jiri Svoboda   Czechoslovakia
2. Jaromir Matena   Czechoslovakia
3. Petko Dinev   Bulgaria
4. Fabrice Pardo   France
5. Ruslan Sharipov   Soviet Union
6. Ralf Glaser   GDR
7. Reinhard Meinel   GDR

II Prizes:

1. Jans Joachim Jetter   FRG
2. Jiri Kolafa   Czechoslovakia
3. Vladimir Reshetov   Soviet Union
4. Grzegorz Zalot   Poland
5. Petr Kučirek   Czechoslovakia
6. Kari Juhani Kujansuu   Finland
7. Josip Lončarić   Yugoslavia
8. Ewald Preiss   FRG

9.	Wolfgang Breymann	FRG
10.	Andrey Ganopolskiy	Soviet Union
11.	Krzysztof Kulpa	Poland
12.	Rupert Leitner	Czechoslovakia
13.	Vladimir Shchukin	Soviet Union
14.	Konstantin G. Tretyachenko	Soviet Union
15.	Mircea A. Vicol	Romania
16.	Galis V. Vedeanu	Romania

III Prizes:

1.	Csaba Biegl	Hungary
2.	Jacek Jasiak	Poland
3.	Alain Parisot	France
4.	Istvan Dory	Hungary
5.	Alexandru Dumitru	Romania
6.	Andreas Bolz	FRG
7.	Björn Danielsson	Sweden
8.	Mathias Hegner	GDR
9.	Jean Louis Jérome	France
10.	György Kriza	Hungary
11.	Olivier Abillon	France
12.	Paweł Grądzki	Poland
13.	Johan Hastad	GDR
14.	Stefan Schuster	GDR
15.	Péter Vankó	Hungary
16.	Jukka Pekka Pekola	Finland

Honourable Mentions:

1.	Mladen Horvatić	Yugoslavia
2.	Zsolt Kovács	Hungary
3.	François Boisson	France
4.	Ove Hanebring	Sweden
5.	Danut Cimpoescu	Romania
6.	Seppo T. Kauntola	Finland

7.	Virgil Bistriceanu	Romania
8.	Ralf Nasilowski	FRG
9.	Mikael Olsson	Sweden
10.	Milen Stefanov Shishkov	Bulgaria
11.	Radoslav Getov	Bulgaria
12.	Jacek Jurczyk	Poland

Best total score (special prize):

     Jiri Svoboda   Czechoslovakia

Special prizes:

* *for the best solution to the problem 1:*

     Ewald Preiss   FRG

* *for the best solution to the problem 2:*

     Andrey Ganopolskiy   Soviet Union

* *for the best solution to the problem 3:*

     Olivier Abillon   France

* *for the best solution to the problem 4:*

     Josip Lončarić   Yugoslavia

* *for the yougest participants:*

     Stefan Schuster   GDR
     Ruslan Sharipov   Soviet Union

XI International Physics Olympiad
Moscow (Soviet Union) 1979

I Prizes:

1.	Maksim Tsipin	Soviet Union
2.	Ivan Ganashev	Bulgaria
3.	Igor Yasonov	Soviet Union
4.	Oleg Yushchuk	Soviet Union
5.	Pavel Kilinaj	Czechoslovakia
6.	Zoltan Kaufman	Hungary
7.	Andrzej Praszmo	Poland
8.	Sergey Shpilkin	Soviet Union

II Prizes:

1.	Bencho Angelov	Bulgaria
2.	Jurgen Grefenstein	GDR
3.	Ion Costin	Romania
4.	Oskar Krenek	Czechoslovakia
5.	Manfred Lehn	FRG

III Prizes:

1.	Gyula Bene	Hungary
2.	Sławomir Drażba	Poland
3.	Cecil Florescu	Romania
4.	Sergey Gordiyenko	Soviet Union
5.	Ventsislav Karadzhov	Bulgaria
6.	Emil Lambrake	Romania
7.	Vladimir Matena	Czechoslovakia
8.	Janusz Oleniacz	Poland
9.	Flavin Pop	Romania
10.	Andreas Ronecker	FRG
11.	Uwe Schmok	FRG
12.	Michał Szypowski	Poland

13. Jan Wohlhein   Czechoslovakia

Honourable Mentions:

  1. Lennart Beierson   Sweden
  2. Dirk Bormann   FRG
  3. Andras Csordas   Hungary
  4. Hartmuth Miks   GDR
  5. Hartmuth Schefer   GDR
  6. Michael Heinrich   GDR
  7. Frank Marlow   GDR
  8. Tiberiu Mokanu   Romania
  9. Wolf-Heinrich Rech   FRG
 10. Ivan Rupski   Hungary
 11. Zoltan Salontai   Hungary
 12. Michał Sośnicki   Poland
 13. Timo Tuuliniemi   Finland
 14. Lars Ulander   Sweden
 15. Ognyan Vasilev   Bulgaria

Best total score: [2]

     Maksim Tsipin   Soviet Union

Special prizes:

* *for the best solution to the problem 2:*

     Lars Ulander   Sweden

----

[2] *The information received thanks to kindness of Dr. Daniel Kluvanec from Czechoslovakia.*

* for the best solutions to the experimental problem:

Hartmuth Miks	GDR
Timo Tuuliniemi	Finland

* for the youngest participant:

Oleg Juszczuk	Soviet Union

The names of the winners were transcibed from the Russian alphabet.

(The names of the winners in each category of prizes are ordered alphabetically)

XII International Physics Olympiad
Varna (Bulgaria) 1981

I Prizes:

1.	Aleksander Gutin	Soviet Union
2.	Andreas Quirrenbach	FRG
3.	Bogdan Tudose	Romania
4.	Iliya Solodovnikov	Soviet Union
5.	Manfred Lehn	FRG
6.	Vladislav Derevyanko	Soviet Union
7.	Wojciech Lerch	Poland

II Prizes:

1.	Jean-François Puget	France
2.	Milan Hanajik	Czechoslovakia
3.	Petr Pavlik	Czechoslovakia
4.	Jurgen Grafenstein	GDR
5.	Andreas Klumper	FRG
6.	Wojciech Kozubski	Poland

7. Adrian Mihai Devenyi	Romania
8. Sandor Palasik	Hungary
9. Andrey Mushinskiy	Soviet Union
10. Milen Penkov	Bulgaria
11. Rossen Varbanov	Bulgaria
12. Silviu Borac	Romania
13. Jiri Matoušek	Czechoslovakia
14. Harald Anlauf	FRG
15. Torbjorn Ledin	Sweden
16. Borivoj Tydliát	Czechoslovakia
17. Grzegorz Szamel	Poland
18. Janos Tamas Pöltl	Hungary
19. Stefan Müller-Pfeiffer	GDR

III Prizes:

1. Grzegorz Majcher	Poland
2. Dag Wedelin	Sweden
3. Erik Heinz	GDR
4. Matiaž Kaluža	Yugoslavia
5. Paweł Trautman	Poland
6. Ovidiu George Radulescu	Romania
7. Miroslav Abrashev	Bulgaria
8. Truong Ba Ha	Vietnam
9. Stefan Müller	FRG
10. Andras Mogyorosi	Hungary
11. Horia Raul Radulescu	Romania
12. Ferenc Gluck	Hungary
13. Dean Možetić	Yugoslavia
14. Matti Airiksinen	Finland
15. Timo Rönkä	Finland
16. Anders Svensson	Sweden
17. Igor Shubyenin	Soviet Union
18. Timo Lattula	Finland

Honourable Mentions:

1.	Ralf Muschall	GDR
2.	Ralf Huonker	GDR
3.	Philipe Rostand	France
4.	Bjorn Ottersten	Sweden
5.	Marc Battyani	France
6.	Jaakko Wegelius	Finland
7.	Heikki Penttila	Finland
8.	Jean-Louis Barrat	France
9.	Le Van Hoang	Vietnam
10.	Petr Plecháč	Czechoslovakia
11.	Jan Carlsson	Sweden
12.	Tikhomir Khristov	Bulgaria
13.	Dimitar Dimitrov	Bulgaria
14.	Tomislav Grčanac	Yugoslavia
15.	Vu Ngoc Tuoc	Vietnam
16.	Ferenc Czako	Hungary

Best total score:

Aleksander Gutin   Soviet Union

Special prizes:

* *for the best solution to the problem 1:*

Stefan Müller   FRG

* *for the best solution to the problem 2:*

Aleksander Gutin   Soviet Union

* *for the best solution to the problem 3:*

Anreas Quirrenbach   FRG

* for the best solution to the experimental problem:

Andrey Mushinskiy   Soviet Union

The names of the winners were transcribed from the Bulgarian alphabet.

XIII International Physics Olympiad
Malente (FRG) 1982

I Prizes:

1. Manfred Lehn   FRG
2. Heikki Tuuri   Finland
3. Adrian Devenyi   Romania
4. Boris Makeyev   Soviet Union
5. Aleksander Filip Żarnecki   Poland
6. Gábor Tóth   Hungary
7. Dariusz Wieczorek   Poland
8. Claudio Emmrich   FRG
9. Michał Kiełkowski   Poland
10. Vladimir Ukhov   Soviet Union
11. Pavel Tsvetkov   Soviet Union
12. Aleksander Panasyuk   Soviet Union
13. Hans—Olof Kuylenstierna   Sweden
14. András Mogyorósi   Hungary

II Prizes:

1. Miroslav Kolesik   Czechoslovakia
2. Jaan Kalda   Soviet Union
3. Zoltán Szállási   Hungary
4. Benjamin Enriquez   France
5. Ovidiu Radulescu   Romania
6. Martin Töx   FRG

7. Johan Sköld — Sweden
8. Djordje Marić — Yugoslavia
9. Jean François Burnol — France
10. Paweł Jałocha — Poland
11. Jiri Bajer — Czechoslovakia
12. Eleodor Nichita — Romania
13. Dean Mozetič — Yugoslavia
14. Dragos Serseni — Romania
15. Niklas Dellby — Sweden
16. Petr Tichavsky — Czechoslovakia
17. Tomislav Grčanac — Yugoslavia
18. Nikolay Neshev — Bulgaria
19. Anders Rantzer — Sweden

III Prizes:

1. Viñh Khanh Nguyen — Vietnam
2. Hartmut Löwen — FRG
3. Hù& Nhañ Hồ — Vietnam
4. Seppo Varho — Finland
5. Duy Thê Trâñ — Vietnam
6. Cezary Juszczak — Poland
7. Marcel Rosu — Romania
8. Daniel Kluvanec — Czechoslovakia
9. Péter Földiák — Hungary
10. Gábor Oszlányi — Czechoslovakia
11. Roman Šášik — Czechoslovakia
12. Stoycho Ivanov — Bulgaria
13. Arend Rensink — The Netherlands
14. Steffen Grossert — GDR
15. Volker Leutheuser — GDR
16. Dimitar Georgiyev — Bulgaria
17. Marc Simon — France
18. Terho Norja — Finland
19. Mikko Kanerva — Finland
20. Torsten Kunz — GDR

21. Mikael Rittri   Sweden
22. Trung Dung Hồ   Vietnam
23. Ralph Huonker   GDR

Honourable Mentions:

1. Ivan Ivanov   Bulgaria
2. Matthias Hübner   GDR
3. Uwe Birkenheuer   FRG
4. Gernot Kunz   Austria
5. Robert Spreeuw   The Netherlands
6. Yves Chateau   France
7. Jyrki Lahtonen   Finland
8. Bruno Rostand   France
9. Matthias Katter   Austria
10. Beloslav Belchev   Bulgaria

Best total score (special prize for the best score and for the third-time-over participation in the International Physics Olympiads):

Manfred Lehn   FRG

Special prizes:

* *for the best solution to the problem 1:*

Dean Mozetič   Yugoslavia

* *for the best solution to the problem 2:*

Aleksander Panasyuk   Soviet Union

* *for the best solution to the problem 3:*

Martin Töx   FRG

* *for the best solution to the problem 4:*

        Dariusz Wieczorek   Poland

* *for the best solution to the problem 5:*

        Aleksander Filip Żarnecki   Poland

* *for the youngest participant:*

        Dean Mozetič   Yugoslavia

XIV International Physics Olympiad
Bucharest (Romania) 1983

I Prizes:

1. Ivan Ivanov   Bulgaria, MD
2. Remus Amilcar Ionescu   Romania, MS
3. Vladimir Viktorovich Molchanov Soviet Union, D
4. Edmond Iancu   Romania, C
5. Andrey Georgiyevich Gnilovskoy Soviet Union, M
6. Anton Yuriyevich Alekseyev   Soviet Union, C
7. Jaroslav Smejkal   Czechoslovakia

II Prizes:

1. Mikhail Borisovich Dyachkov   Soviet Union, MO
2. Marius Goldenberg   Romania, MS
3. Krzysztof Mnich   Poland, S
4. Erik Hakvoort   The Netherlands,O
5. Paweł Jałocha   Poland
6. Roland Schultze   FRG
7. Thomas Kerler   FRG
8. Marius Vasiliu   Romania, C

9. Daniel Kluvanec   Czechoslovakia

III Prizes:

1. Jean-Michel Courty   France, E
2. Valdis Yurevich Birzvalks   Soviet Union
3. Steffen Grossert   GDR
4. Janos Fent   Hungary
5. Naray Miklos   Hungary
6. Dorel Moldovan   Romania
7. Darko Stefanovic   Yugoslavia
8. Zbigniew Płuciennik   Poland
9. Milos Drutacovsky   Czechoslovakia
10. Arkossy Otto   Hungary
11. Gyuricza Béla   Hungary
12. Nguyen Quang Son   Vietnam, M
13. Andrea Hölschner   FRG
14. Stoytsyo Ivanov   Bulgaria
15. Robert Sandrock   FRG
16. Wulf Böttiger   GDR

Honourable Mentions:

1. Daniel Poloni   FRG
2. Phan Thanh Hai   Vietnam
3. Tran Hun Huan   Vietnam
4. Frei Zsolt   Hungary
5. Viktor Ivanov   Bulgaria
6. Antero Hietamaki   Finland
7. Ivo Koren   Czechoslovakia
8. Anders Eriksson   Sweden
9. Roman Sasik   Czechoslovakia
10. Klaus Aufinger   Austria
11. Igor Herbert   Yugoslavia
12. Michel Arboi   France
13. Mans Henningson   Sweden

14. Guillaume Aubin   France
15 Jean Louis Dufour   France
16. Lauri Pirttiaho   Finland
17. Anders Lunblad   Sweden

Best total score (special prize):

    Ivan Ivanov   Bulgaria

Special prizes have been denoted, following names of the winners, by means of the following codes:

    M − *special prize for the problem 1*
    E − *special prize for the problem 2*
    O − *special prize for the problem 3*
    C − *special prize for the problem 4*
    S − *special prize for the additional*
                                   non-obligatoty problem
    D − *special prize for the experimental problem*

The special prizes were awarded also to the following participants:

    1. Nikola Petrov   Bulgaria, M
    2. Herve Bercegol   France, C

XV International Physics Olympiad
Sigtuna (Sweden) 1984

I Prizes:

    1. Anton Alekseyev   Soviet Union
    2. Ian de Boer   The Netherlands
    3. Aleksander Dyeshkovskiy   Soviet Union
    4. Fodor Gyula   Hungary
    5. Jan Lużny   Czechoslovakia

6.	Sergey Orlov	Soviet Union
7.	Dan Pirjol	Romania
8.	Sorin Spânoche	Romania
9.	Marius Vasiliu	Romania

II Prizes:

1.	Dirk Graudenz	FRG
2.	Thomas Müller	FRG
3.	Igor Potyeryayko	Soviet Union
4.	Mihail Yotov	Bulgaria
5.	Lev Zakshevskiy	Soviet Union

III Prizes:

1.	Klaus Becker	FRG
2.	Peter Csillag	Hungary
3.	James Durrant	Great Britain
4.	Laszlo Erdös	Hungary
5.	Jozsef Frigo	Hungary
6.	Viktor Ivanov	Bulgaria
7.	Olaf Kalz	GDR
8.	Cristian Neascu	Romania
9.	Dietmar Polster	GDR
10.	Frank Robijn	The Netherlands
11.	Viliam Schichman	Czechoslovakia
12.	Paul Shutler	Great Britain
13.	Michael Skeide	FRG
14.	Vasil Spasov	Bulgaria
15.	Cristian Teodorescu	Romania
16.	Wojciech Zabołotny	Poland

Honourable Mentions:

1.	Gerard Barkema	The Netherlands
2.	Jarosław Bogusz	Poland

3.	Wulf Böttger	GDR
4.	Anders Eriksson	Sweden
5.	Gabor Fath	Hungary
6.	David Griffith	Great Britain
7.	Jyrki Kivinen	Finland
8.	Dagmar Kluvancová	Czechoslovakia
9.	Klaus Krischan	Austria
10.	Pieter Maris	The Netherlands
11.	Josef Pelikán	Czechoslovakia
12.	Michael Ringe	FRG
13.	Ivan Sirakov	Bulgaria
14.	Patrik Spanel	Czechoslovakia
15.	Adam Strzeboński	Poland
16.	Menke Ubbens	The Netherlands
17.	Matjaz Zitnik	Yugoslavia

Best total scores (ex aequo, special prizes):

Ian de Boer	The Netherlands
Sorin Spădoche	Romania

Special prizes:

* *for the best total score in theory:*

Ian de Boer	The Netherlands

* *for the best total score in experiment:*

Marius Vasiliu	Romania

* *for the most original solution:*

Anton Alekseyev	Soviet Union

* *for the best formal presentation of the solutions to the teoretical problems:*

        Jan Lużny   Czechoslovakia

* *for the best formal presentation of the solutions to the experimental problems:*

        Dirk Graudenz   FRG

* *for the youngest participant:*

        Lev Zakshevskiy   Soviet Union

* *for the best participants from very far countries:*

        Alberto Clavijo   Cuba
        Tran Nhat Quang   Vietnam

* *for the best participants from the countries taking part in the competition for the first time:*

        Paul Shutler   Great Britain
        Vilhjálmur Thorsteinson   Iceland
        Lars Hoff   Norway

* *for the best score by a female participant:*

        Dagmar Kluvancová   Czechoslovakia

* *for the second best score by a female participant:*

        Anna Puntajer   Austria

XVI International Physics Olympiad
Portorož (Yugoslavia) 1985

I Prizes:

1. Roy Badami   Great Britain
2. Viktor Bazhykin   Soviet Union
3. Georgiy Grigoryev   Soviet Union
4. Taras Ivanenko   Soviet Union
5. Patrik Spanel   Czechoslovakia

II Prizes:

1. Norbert Bollow   FRG
2. Zoltan Egyed   Hungary
3. David Mackay   Great Britain
4. Dan Przzol   Romania
5. Peter Schupp   FRG
6. Yuriy Zhestkov   Soviet Union

III Prizes:

1. Phons Bloemen   The Netherlands
2. Oleg Cherp   Soviet Union
3. Igor Djoković   Yugoslavia
4. Anthony Duell   Great Britain
5. Richard Green   Great Britain
6. Reiner Hippmann   FRG
7. Antal Jakovac   Hungary
8. Mathias Ketzel   GDR
9. Ovidiu Klocea   Romania
10. Jane Kondev   Yugoslavia
11. Mirosław Lis   Poland
12. Jan Lużny   Czechoslovakia
13. Katalin Malureanu   Romania
14. Thomas Palm   Sweden

15.	Olaf Wendt	FRG
16.	Marcin Wolter	Poland
17.	Peter Zegelaar	The Netherlands

Honourable Mentions:

1.	Lars Aronsson	Sweden
2.	Nicolas Bateman	Canada
3.	Dobrin Bosev	Bulgaria
4.	Mathias Drochner	GDR
5.	Jari-Pekka Ikonen	Finland
6.	Henrik Jurkschat	GDR
7.	Thomas Klotz	GDR
8.	Stefan Komilev	Bulgaria
9.	Nikolay Mechkov	Bulgaria
10.	Ivo Myslivec	Czechoslovakia
11.	Kristian Neasu	Romania
12.	Akos Nemeth-Buhin	Hungary
13.	Jeroen Nijhof	The Netherlands
14.	Nguyen Ninh Khang	Vietnam
15.	Veikko Punkka	Finland
16.	Jens-Uwe Sachse	GDR
17.	Jörg Schwelberger	Austria
18.	Przemysław Siemion	Poland
19.	Torbjorn Soderberg	Sweden
20.	Harun Solak	Turkey
21.	Håkan Svensson	Sweden
22.	Ralf Vendenhouten	FRG
23.	Veli-Rekka Viitanen	Finland
24.	Jacek Wójcik	Poland
25.	Phan Xuan Hai	Vietnam

Best total score (special prize):

Patrik Spanel	Czechoslovakia

Special prizes:

* *for the best total score in theory:*

        Taras Ivanenko   Soviet Union

* *for the best total score in experiment:*

        David Mackay   Great Britain

* *for the youngest participant:*

        Viktor Bazhykin   Soviet Union

* *for the best participants from the countries taking part in the competition for the first time:*

        Nicolas Bateman   The Netherlands
        Harun Solak   Turkey

* *for the most humorous solution:*

        Lars Aronsson   Sweden

(The names of the winners in each category of prizes are ordered alphabetically)

XVII International Physics Olympiad
Londyn-Harrow (Great Britain) 1986

I Prizes:

1. O. Volkov   Soviet Union
2. S. Myachilov   Soviet Union
3. C. Necula   Romania
4. A. Matychin   Soviet Union

II Prizes:

1. M. Pearlman   Great Britain
2. A. Gushchin   Soviet Union
3. A. Maasen van der Brink The Netherlands
4. A. Bot   Romania
5. Lin Chen   China

III Prizes:

1. Z. Kantor   Hungary
2. J. Wójcik   Poland
3. B. de Backer   The Netherlands
4. D. Vrinceanu   Romania
5. A. Jakovac   Hungary
6. P. Kolnik   Czechoslovakia
7. R. Ledrinka   Czechoslovakia
8. P. Schupp   FRG
9. P. Graham   USA
10. N. Ketzel   GDR
11. Z. Gagyi-Palffy   Romania
12. Wei Xing   China
13. M. Andrews   Great Britain
14. J. Schuth   FRG
15. A. Kaiser   Hungary
16. J. Zucker   USA
17. F. Fischer   FRG
18. P. Mauskoff   USA
19. G. Nikolaishvili   Soviet Union
20. D. Joyce   Great Britain
21. W. Skiba   Poland
22. M. P. Ertas   Turkey

Honourable Mentions:

1.	M.J.Nyhof	The Netherlands
2.	Zang Ming	China
3.	P. Siemion	Poland
4.	P. Peeb	Bulgaria
5.	F. Klemm	GDR
6.	A. Verberkmoes	The Netherlands
7.	T. Klotz	GDR
8.	A. Leitereg	Hungary
9.	D. Aarvold	Great Britain
10.	R. Weise	FRG
11.	R. Adamec	Czechoslovakia
12.	M. Hoffmann	GDR
13.	H. Held	FRG
14.	K. Asotrom	Sweden
15.	S. Marksteiner	Austria
16.	P. Jastrzębski	Poland
17.	L. Beskow	Sweden
18.	T. Tasnadi	Hungary
19.	D. Phalp	Canada
20.	N. Craig-Wood	Great Britain
21.	S. Gaind	Canada
22.	S. Bossev	Bulgaria
23.	R. Chrzan	Poland
24.	B. Vulpescu	Romania
25.	J. Laaksonen	Finland
26.	N. Mechkov	Bulgaria
27.	S. Meduna	Czechoslovakia

Best total score (special prize):

       O. Volkov   *Soviet Union*

Special prizes:

* *for the best solution to the problem 1:*

      P. Kolnik   Czechoslovakia

* *for the best solution to the problem 2:*

      S. Marksteiner   Austria

* *for the best solution to the problem 3:*

      J. Schuth   FRG
      Z. Gagyi-Palffy   Romania

* *for the best performance of the experiment on the rainbow:*

      S. Myachilov   Soviet Union

* *for the best solution to the computer problem:*

      M. Pearlman   Great Britain

* *for the best score by a female participant:*

      Lotte Beskow   Sweden

XVIII International Physics Olympiad
Jena (GDR) 1987

I Prizes:

1. Catalin Malureanu   Romania
2. Ciprian Necula   Romania
3. Bastiaan V. de Bakker   The Netherlands

II Prizes:

1.	Chen Xun	China
2.	Klaus Hollatschek	FRG
3.	Didina Serban	Romania
4.	Cristoph Tiele	FRG
5.	Christopher Sanders	Great Britain
6.	Bogdan Vulpescu	Romania
7.	Anton Bibikov	Soviet Union
8.	Li Jin Hui	China
9.	Frank Klemm	GDR
10.	Gabor Drasny	Hungary

III Prizes:

1.	Piotr Kossacki	Poland
2.	Gunther Seitz	FRG
3.	Jürgen Hertzfeld	GDR
4.	Pavel Kolnik	Czechoslovakia
5.	Gyula Szokoly	Hungary
6.	Bryan Beatty	USA
7.	Andrzej Pyka	Poland
8.	Boris Karadzov	Bulgaria
9.	Martin Krause	FRG
10.	Arthur Hebecker	GDR
11.	Steffen Winterfeld	GDR
12.	Lucian Mitoseriu	Romania
13.	Patrik Pettersson	Sweden
14.	Eli Glezer	USA
15.	Petar Maksimović	Yugoslavia
16.	Ivan Georiyev Tsvetanov	Bulgaria
17.	Karl Berggren	Canada
18.	Gergely Zaránd	Hungary
19.	Wilold Skiba	Poland
20.	Tang Peng Fei	China
21.	Normand Modine	USA

22. Ho Si Mau Thuc	Vietnam
23. Andrej Vilfan	Yugoslavia
24. Jaroslav Hora	Czechoslovakia
25. Robbert R. R. Nix	The Netherlands
26. Dimitriy Budko	Soviet Union
27. Aleksey Goldin	Soviet Union
28. Wu Aj Hua	China
29. Zhang Yan Ping	China

Honourable Mentions:

1. Stephan Jeiter	FRG
2. David Maxera	Czechoslovakia
3. Martin Andrews	Great Britain
4. William Christopherson	Great Britain
5. Martin Erikson	Sweden
6. Jari-Pekka Paalassalo	Finland
7. Gintas Vilkyalis	Soviet Union
8. Dmitriy Glushchenkov	Soviet Union
9. Karl Strobl	Austria
10. András Vasy	Hungary
11. Grzegorz Kondrat	Poland
12. Gregory Wellman	Canada
13. Dirk K. de Vries	The Netherlands
14. Pham Hung	Vietnam
15. Walter Renger	Austria
16. Peter Balogh	Hungary
17. Marcin Konecki	Poland
18. Tomas Eriksson	Sweden
19. Håkan Hjalmers	Sweden
20. Öjvind Johansson	Sweden
21. Robert Gassler	Austria
22. Truong Dinh Ngo Quang	Vietnam
23. Katarina Kis. Petrikova	Czechoslovakia
24. Hervé Kuijten	The Netherlands
25. Timothy Creasy	Canada

26.	Petr Habala	Czechoslovakia
27.	Gunnar Gudnason	Iceland
28.	Panu Lehtovuori	Finland
29.	Franklin Chen	USA
30.	Nguyen Son Tung	Vietnam

Best total score (special prize):

Catalin Malureanu   Rumunia

Special prizes:

* for the best total score in theory:

Catalin Malureanu   Romania
Anton Bibikov   Soviet Union

* for the best score by a female participant:

Didina Serban   Romania

XIX International Physics Olympiad
Bad Ischl (Austria) 1988

I Prizes:

1.	Conrad McDonnell	Great Britain
2.	Gabor Drasny	Hungary
3.	Gunther Seitz	FRG
4.	Yangsong Chen	China
5.	Boris Karadyov	Bulgaria
6.	Klaus Hallatschek	FRG
7.	Costin Popescu	Romania

II Prizes:

1.	Yuri Kravchenko	Soviet Union
2.	Christian Gavrila	Romania
3.	Duiliu Diaconescu	Romania
4.	Mats Persson	Sweden
5.	Michael Edwards	USA
6.	Stefan Simion	Romania
7.	Cezary Śliwa	Poland
8.	Jianbo Xu	China
9.	Norbert Miskolczi	Hungary
10.	Matthew Stone	USA
11.	Piotr Kossacki	Poland
12.	Rashmi Tank	Great Britain
13.	Kostya Penanen	Soviet Union
14.	Radek Vystavel	Czechoslovakia
15.	Aleksander Mazurenko	Soviet Union
16.	Feng Chen	China
17.	Hans Olsson	Sweden
18.	Lucian Ciobica	Romania
19.	Daniel Bertilsson	Sweden
20.	Ian Lovejoy	USA
21.	Reyer Gerlagh	The Netherlands
22.	Ivan Popov	Bulgaria
23.	Tamas Hauer	Hungary

III Prizes:

1.	Pierpaolo Peirano	Italy
2.	Kari Nurmela	Finland
3.	Sergey Cheshkov	Bulgaria
4.	Vadim Moroz	Soviet Union
5.	David Jackson	Australia
6.	Mattias Markert	GDR
7.	Trevor Blackwell	Canada
8.	Erkki Lantto	Finland

9.	Anton Malkim	Soviet Union
10.	Tibor Bartos	Czechoslovakia
11.	Andrej Vilfan	Yugoslavia
12.	Wolfgang Maichen	Austria
13.	Zvonimir Bandic	Yugoslavia
14.	Thomas Koch	GDR
15.	Jeroen Paasschens	The Netherlands
16.	Stuart Yates	Australia
17.	Bernd Eberth Ammann	FRG
18.	Thomas Boeck	GDR
19.	Simeon Stoyanov	Bulgaria
20.	Udo Karthaus	FRG
21.	Csaba Csaki	Hungary
22.	Mark Gorbatov	Australia
23.	Attila Fucsar	Hungary
24.	Aidong Ding	China
25.	Roar Lauritzen	Norway
26.	Christopher Balzereit	FRG
27.	Neil Greenham	Great Britain
28.	David Maxera	Czechoslovakia
29.	Marko Santala	Finland

Honourable Mentions:

1.	Gunnar Farnebaeck	Sweden
2.	Thomas Wilcke	GDR
3.	Grzegorz Kondrat	Poland
4.	Kees Van Kemenade	The Netherlands
5.	Jan Swart	The Netherlands
6.	Lennart Bengtsson	Sweden
7.	Katarina Kis-Petrikova	Czechoslovakia
8.	Jian Chen	China
9.	Maciej Sawicki	Poland
10.	Tuan Phan	Vietnam
11.	Dietmar Kieslinger	Austria
12.	Leopoldo A. Pando Zayas	Cuba

13.	David Hogg	Canada
14.	Georgios Hadjiyannis	Cyprus
15.	Arne H. Juul	Norway
16.	Mauro Cossi	Italy
17.	Hakon Asgrimsson	Iceland
18.	Gregor Weihs	Austria
19.	Sjoerd Stallinga	The Netherlands
20.	Timothy Little	Australia
21.	Hoyt Hudson	USA
22.	Mikko Laine	Finland
23.	Rossen Ivanov	Bulgaria
24.	James Gifford	Australia
25.	Trond Reitan	Norway
26.	Lars B. Schroeder	Norway
27.	Patrik Rautaheimo	Finland

Best total score (special prize):

Conrad McDonnell   Great Britain

Special prizes:

* for the best solution to the problem 1:

Aidong Ding   China

* for the best solution to the problem 2:

Jan Swart   The Netherlands

* for the best solution to the problem 3:

Thomas Wilcke   GDR

for the best solution to the problem 4:

        Piotr Kossacki   Poland

for the youngest participant:

        Cezary Śliwa   Poland

XX International Physics Olympiad
Warszawa (Poland) 1989

I Prizes:

1. Steven Gubser   USA
2. Szabolcs Kèsmàrki   Hungary
3. Costin-Radu Popescu   Romania
4. Olaf Kummer   FRG
5. Jens Lang   FRG
6. Michael Rutter   Great Britain
7. Desmond Rodney Lim Chin Siong   Singapore
8. Nikolay Kuzma   Soviet Union
9. Asen Kumanov   Bulgaria
10. Eric Cator   The Netherlands

II Prizes:

1. Cezary Śliwa   Poland
2. Piotr Kossacki   Poland
3. Yan Jing   China
4. Tomasz Motylewski   Poland
5. Andrej Vilfan   Yugoslavia
6. Mao Yong   China
7. Udo Karthaus   FRG
8. Colin Merryweather   Great Britain
9. Andre Fraenzel   GDR
10. Gabriel Balan   Romania

11.	Gregory Colyer	Great Britain
12.	Qiu Dong Yu	China
13.	Ge Ning	China
14.	Mika Nyström	Sweden
15.	Rolf Oldeman	The Netherlands
16.	Szilàrd Szabò	Hungary
17.	Arnošt Kobylka	Czechoslovakia
18.	Volker Gebhardt	FRG
19.	Gabòr Felsö	Hungary
20.	Leopoldo Avelino Pando Zayas	Cuba
21.	Romke Jonker	The Netherlands
22.	Gregor Weihs	Austria
23.	Aleksandr Korshtsov	Soviet Union
24.	Miroslav Vicher	Czechoslovakia
25.	Konstantin Stefanov	Bulgaria
26.	Werner Torsten	GDR

III Prizes:

1.	Arthur Street	Australia
2.	Lim Shiang Liang	Singapore
3.	Zvonimir Bandić	Yugoslavia
4.	Thomas Wilcke	GDR
5.	Konstantin Zuev	Soviet Union
6.	Nima Arkani-Hamed	Canada
7.	Richard Wilson	Great Britain
8.	Stefan Jacobsson	Sweden
9.	Øyvind Tafjord	Norway
10.	Andrej Doboš	Czechoslovakia
11.	Jason Jacobs	USA
12.	Lucian Ciobica	Romania
13.	Romuald Janik	Poland
14.	Swen Wunderlich	GDR
15.	Volker Springel	FRG
16.	Yavor Velchev	Bulgaria
17.	Ramin Farjad Rad	Iran

18. Zoltàn Hidvègi	Hungary
19. Lin Xiao Fan	China
20. Viorel-Cristian Negoita	Romania
21. Carsten Deus	GDR
22. Hannes Sakolin	Austria
23. Ramin Golestanian	Iran
24. Matthew Brecknell	Australia
25. Martijn Mulders	The Netherlands
26. Thomas Bednar	Austria
27. Dubravko Tomasović	Yugoslavia
28. Derrick Bass	USA
29. Hans Olav Sundfør	Norway
30. Chris Simons	Canada

Honourable Mentions:

1. Vahid Borumand Sani	Iran
2. Vladislav Makeyev	Soviet Union
3. Mona Berciu	Romania
4. Timo Tarhasaari	Finland
5. Silvano de Franceschi	Italy
6. Dalibor Tużinski	Yugoslavia
7. Federico Toschi	Italy
8. Gareth Williams	Australia
9. Christophe Colle	Belgium
10. James Sarvis	USA
11. Hüseyin Altun	Turkey
12. Timo Rantalainen	Finland
13. Norbert Schörghofer	Austria
14. Georgios Ioannou	Cyprus
15. Janne Karimäki	Finland
16. Brett Munro	Australia
17. Gregory Lielens	Belgium
18. Petr Duczynski	Czechoslovakia
19. Simon Ekström	Sweden
20. Gordon Ogilvie	Great Britain

21.	Johan Axnäs	Sweden
22.	Pekka Heino	Finland
23.	Eric Nodwell	Canada
24.	Roger Klausen	Norway
25.	Stefan Piperov	Bulgaria
26.	Kristjàn Leòsson	Iceland
27.	Gerardo A. Muñoz	Columbia
28.	Otso Ovaskainen	Finland
29.	Juru Uvarov	Soviet Union
30.	Dragomir Nechev	Bulgaria
31.	Leszek Mencnarowski	Poland
32.	Arnold Metselaar	The Netherlands
33.	Vjekoslav Mladineo	Yugoslavia

Best total score (special prize):

Steven Gubser   USA

Special prizes:

* *for the the second highest score:*

Szabolcs Kèsmàrki   Hungary

* *for the best solution to the problem 1:*

Desmond Roney Lim Chin Siong   Singapore

* *for the best solution to the problem 2:*

Gabriel Balan   Romania

* *for the best solution to the problem 3:*

Ge Ning   China

* *for the best solution to the experimental problem:*

Romke Jonker	Holland
Colin Merryweather	Great Britain

* *for the highest score by a female participant:*

Mona Berciu	Romania

* *for the youngest participant:*

Jessica Millar	USA

* *for the best combined score to the problem 1 and the experimental problem (founded by the Institute of Low Temperatures and Structural Research, Polish Academy of Sciences, Wrocław, Poland):*

Colin Merryweather	Great Britain

* *for the best balanced experimental and theoretical solutions (found by the European Physical Society):*

Andrej Doboš	Czechoslovakia

* *for the most original solution (found by Professor I. Białynicki-Birula, the winner of the 1st Polish Physics Olympiad):*

Gabriel Balan	Romania

STATISTICS

N A T I O N A L   P H Y S I C S   O L Y M P I A D S
(Comparison)

Austria	3CD (450e, 160e, 24e. 5) [5 part., 3 days, theor. only, yes]	P
Bulgaria	4GD (50000, 15000, 1000, 60e, 10) [10 part., 20-30 days, exp.+theor., yes]	MP
Canada	3CD (500, 65e, 12e, 12) [12 part., 5 days, exp.+theor., yes]	PS
China	2CD (56000e, 100e, ?) [15 part., 60 days, exp.+theor., yes]	MPS
Colombia	3GD (4500, 500, 50e, ?) [30 part., 45 days, exp.+theor., yes]	PS
Cyprus	3CS (200, 35, 10e, ?) [10 part., 50 days, exp.+theor., yes]	–
Finland	3GS (700, 1000, 30e, ?) [5 part., 4 days, exp. only, yes]	S
German D. R.	2GS (600, 35e, 5) [6 part., 15 days, exp.+theor., yes]	P
Great Britain	3CS (20000, 500, 20e. 20) [no training]	P
Hungary	3GD (6000, 600, 50e, ?) [60 part., 15 days, exp.+theor., yes]	M
I. R. Iran	3CS (4000, 50e, 7e. 5) [7 part., 50 days, exp.+theor., yes]	MP
Italy	3CS (3000, 400, 50, 10) [10 part., 5 days, exp.+theor., yes]	P
The Netherlands	2CS (1100, 20e, 5) [5 part., 5 days, exp. only, no]	P
Norway	2CS (400, 50, 8) [5 part., 5 days, exp.+theor., no]	P
Poland	4CS (200Ce, 1500e, 1000t/400e, 80, 25) [10 part., 10 days, exp.+theor., yes]	MP
Sweden	2CS (1000, 15e, ?) [5 part., 3 days, exp.+theor., no]	PS
Turkey	3GS (120, 80, 20e, ?) [17-20 part., 45 days, exp.+theor., yes]	P

→

126

Explanation:

First number: number of stages

Following letters: C – Common competition for all the
                   grades
                   G – the competition is organized for
                   each Grade separetely
                   D – the problems can be Different for
                   different participants
                   S – the Same problems for all the
                   participants

in brackets: approximate numbers of participants in
                   subsequent stages of the competition and
                   the number of winners ("e" after a
                   number means that an experimental
                   problem is given at the stage
                   corresponding to the number)

following letters: privileges:
                   M – Merit when applying for the
                   universities
                   P – Prize (money or gifts)
                   S – Scholarship

next row in "[..]": data cocerning the training [number of
                   participants. length of the training in
                   days, character of the training. does
                   any final selection of pupils depend on
                   their activity during the training?]

DISTRIBUTION OF PRIZES IN TWENTY INTERNATIONAL PHYSICS OLYMPIADS

Country	PV	I(A)	II	III	HM	Σ	Σ/PV
Australia	15	–	–	5	4	9	0.60
Austria	40	–	1	3	12	16	0.40
Belgium	10	–	–	–	2	2	0.20
Bulgaria	97	7(1)	8	18	28	61	0.63
Canada	25	–	–	4	7	11	0.44
P. R. China	20	1	9	6	2	18	0.90
Colombia	10	–	–	–	1	1	0.10
Cuba	40	–	1	–	1	2	0.05
Cyprus	10	–	–	–	2	2	0.20
Czechoslovakia	97	14(4)	21	24	23	82	0.85
Finland	55	1	1	10	21	33	0.60
France	35	2	5	9	14	30	0.86
German D. R.	94	7	12	34	26	79	0.84
F. R. Germany	70	8(1)	17	23	11	59	0.84
Great Britain	30	3(1)	6	8	6	23	0.77
Greece	5	–	–	–	–	0	0.00
Hungary	97	16(2)	17	36	19	88	0.91
Iceland	30	–	–	–	3	3	0.10
I. R. Iran	5	–	–	2	1	3	0.60
Italy	25	–	–	1	3	4	0.16
Kuwait	15	–	–	–	–	0	0.00
The Netherlands	40	3(1)	5	8	13	29	0.73
Norway	30	–	–	3	4	7	0.23
Poland	97	13(4)	18	27	26	84	0.87
Romania	97	15(2)	24	29	14	82	0.85
Singapore	5	1	–	1	–	2	0.40
Soviet Union	94	37(5)	23	20	12	92	0.98
Sweden	60	1	8	8	23	40	0.67
Turkey	20	–	–	1	2	3	0.15
USA	20	1(1)	3	8	3	15	0.75
Vietnam	35	–	–	7	10	17	0.49
Yugoslavia	69	1	7	14	10	32	0.46
TOTAL	1392	131	186	309	303	929	
Lithuanian SSR	5	–	–	1	1	2	0.40
TOTAL[*]	1397	131	186	310	304	931	

[*] including the Lithuanian SSR (unofficial participation)

→

Explanation:

PV — "Participation Volume", i.e. the total possible number of participants (from a given country) in the Olympiads in which the country took part

I(A) — number of the **I** prize winners (in brackets: number of the Absolute Winners)

II — number of the **II** prize winners

III — number of the **III** prize winners

HM — number of the participants who obtained the Honourable Mentions

Σ — number of the participants who passed the competition successfully (I+II+III+HM)

Σ/PV — number of successful passes per one participant and per one Olympiad

COMMENTS:

1. In the 1st IPhO there were 7 prize winners classified linearly. As their scores were very high and very close to each other, in the above Table all the winners of the 1st IPhO are treated as the I prize winners.

2. The total number of the Absolute Winners is 22 instead of 20 as in two Olympiads two students were classified ex aequo in the first position.

3. In the case of a country sending always a full team the value of PV equals to the total number of its participants in the Olympiads in which the country took part. Some countries, however, send fewer pupils than it is allowed. Then, of course, the real total numbers of their participants are smaller than the corresponding values of PV. Unfortunately, the list of the prize winners (included in this issue) does not allow for reconstruction of the real total number of participants from each country since it does not include those who have not obtained any prize or honourable mention.

N U M B E R S   O F   P R I Z E S   I N   S U B S E Q U E N T   I N T E R N A T I O N A L   P H Y S I C S   O L Y M P I A D S

Olympiad	P	C	I	II	III	HM	S	Σ	VO	Σ/VO
I	3	5	7	←*)	←*)	4	–	11	15	0.73
II	3	8	2	1	3	2	1	8	24	0.33
III	5	8	13	10	9	5	–	37	40	0.93
IV	6	8	4	7	10	13	2	34	48	0.71
V	5	7	5	6	12	7	–	30	35	0.86
VI	5	9	2	3	14	15	7	34	45	0.76
VII	5	8	3	–	8	7	2	18	40	0.45
VIII	5	9	7	9	12	8	1	36	45	0.80
IX	5	10	7	7	12	13	1	39	50	0.78
X	5	12	7	16	16	12	7	51	60	0.85
XI	5	11	8	5	13	15	4	41	55	0.75
XII	5	14	7	19	18	16	4	60	70	0.86
XIII	5	17	14	19	23	10	7	66	85	0.78
XIV	5	16	7	9	16	17	19	49	80	0.61
XV	5	18	9	5	16	17	15	47	90	0.52
XVI	5	20	5	6	17	25	7	53	100	0.53
XVII	5	21	4	5	22	27	8	58	105	0.55
XVIII	5	25	3	10	29	30	4	72	125	0.58
XIX	5	27	7	23	29	27	6	86	135	0.64
XX	5	29	10	26	30	33	12	99	145	0.68
XX**)	5	30	10	26	31	34	12	101	150	0.67
TOTAL			131	185	309	303		929	1392	
TOTAL**)			131	186	310	304		931	1397	
mean value										0.67

----

*) see the comments to the previous Table   →

**) including the Lithuanian SSR

Explanation:

P   — alowed number of Pupils in each team
C   — number of participating Countries
I   — number of the I prize winners
II — number of the II prize winners
III — number of the III prize winners
HM — number of the participants who obtained the
      Honorouble Mentions
S   — number of Special prizes awarded
Σ   — number of the participants who passed the competition
      successfully (I+II+III+HM)
VO — "Volume of the Olympiad", i.e. the total possible
      number of participants in a given competition (P×C)
Σ/VO — number of successful passes per one participant in a
       given Olympiad

COMMENT:

    VO is practically equal to the total number of
participants in a given Olympiad since usually all the
countries send full teams to the competition. (Few
exceptions, however, have happened sometimes.) Thus, the
quantity Σ/OV is the number of successful passes per one
participant in a given Olympiad.

P R O B L E M S   A N D   T H E I R   M A R K I N G

Olympiad	C	T	E	MT	ME	M	MS	MS/M
I	L	4*)	1	10.10.10.6	10	40	39	98
II	A	3	1	10,10,10	10	40	35	88
III	A	4	1	8,8,8,8	16	48	48	100
IV#)	A	4	1	10,10,10,10	20	60	57	95
V#)	A	4	1	10,10,10,10	20	60	48.6	81
VI#)	A†)	4	1	10,10,10,10	20	60	57	95
VII	R	3	1	10,10,10	20	50	46	92
VIII	R	3	1	10,10,10	20	50	43	86
IX	R	3	1	10,10,10	20	50	47.5	95
X	R	3	1	10,10,10	20	50	49	98
XI	R	3	1	10,10,10	20	50	43	86
XII	R	3	1	10,10,10	20	50	47	94
XIII	R	3	2	10,10,10	10,10	50	43	86
XIV	R	4+S‡)	1	8,8,7,7	20	50	43.75	87.5
XV	R	3	2	10,10,10	10,10	50	43	86
XVI	R	3	2	10,10,10	10,10	50	42.5	85
XVII	R	3	2	10,10,10	10,10	50	37.9	75.8
XVIII	R	3	1	10,10,10	20	50	49	98
XIX	R	3	2	10,10,10	10,10	50	39.38	78.8
XX	R3	3	1	10,10,10	20	50	46.333	92.7

----

*) the pupils were required to solve three theoretical problems (of their choice)

†) all the results were multiplied by 1.2 prior to awarding the prizes (in the absolute scale the best score was 47.5 points; it corresponds to MS/M = 79.2 %)

‡) the pupils got four "regular" problems and one additional problem (for special prize only)

#) the data concerning the IV. V and VI IPhOs were reconstructed due to kind help of Ms. G. S. Tarasyuk

→

Explanation:

C   — Classification system:

       L   — linear ordering of the winners according to the total number of points obtained

       A   — three groups of the prize winners and one group of those who obtained the honourable mentions — the minima for each group were calculated in respect to the theoretically possible highest score

       R   — as A, but the minima were calculated in respect to the best score reached at a given Olympiad

       R3 — as A, but the minima were calculated in respect to the mean value of the three best scores reached at a given Olympiad

T   — number of the Theoretical problems

E   — number of the Experimental problems

MT   — possible Maximum scores for each of the Theoretical problems

ME   — possible Maximum scores for each of the Experimental problems

M   — theoretically possible Maximum total score at a given Olympiad

MS   — Maximum Score reached at a given Olympiad

MS/M — the ration (in %) of MS to M

*Some statistical data are contained also in the Tables
I and II in the article on the history of the IPhOs.*

AUSTRIA

AUSTRIAN PHYSICS OLYMPIAD

Prof. Günther LECHNER   Prof. Helmuth MAYR

PROCEDURE FOR SELECTING TEAMS TO THE INTERNATIONAL PHYSICS OLYMPIADS

All students of secondary schools are allowed to attend an optional course of physics at school in addition to physics as a compulsory subject.

Such courses take place once a week and take two lessons. These lessons start in October and end in April. By the end of April the national competition starts:

There are three levels, the so-called "Kurswett= bewerb", the "Landeswettbewerb" and the "Bundes= wettbewerb".

The "Kurswettbewerb" (= course-competition) is a competition of the students of one course.

The teacher of the individual course is responsible for the theoretical and experimental problems of this competition.

It takes three hours.

Those four students of each course in one of the nine federal-lands of Austria who have proved to be best in their course-competition may take part in a "Landeswettbewerb" (= federal-land competition).

All members of these federal-land competitions have to solve the same theoretical and experimental problems at the same time.
There are four hours to solve the problems.

The best students from these federal-land competit= ion are qualified to take part in the "Bundeswett= bewerb" (= federal-competition).

There is a ten-days course before the federal-competition.
About 24 students take part in it. After the course the two-day competition follows: On the first day there is a five-hour theoretical competition, on the second day there is a five-hour experimental competition.

The five students with the best qualification of this federal competition are allowed to participate in the IPHO.
They take part in a special training. It takes three days.

Experimental Problem (ÖPHO 1984)

To measure you can use:
1 oszillating (solid) body
1 stopwatch
1 tape rule
1 piece of thread
2 weights
2 cylinders of metal (m = ... g; D = ... mm)
1 Chinese-ink-pencil
1 knitting-needle
   glue
   labels
   millimeter-ruled paper
   tripod-material

moment of inertia of a cylinder:

$$I = \frac{1}{2}.m.r^2$$

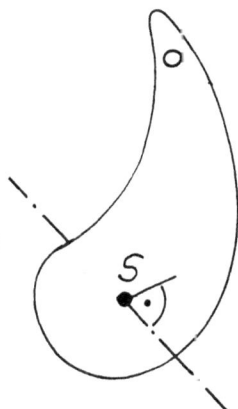

1. Determine the moment of inertia of the given (solid) body relative to an axle orthogonal to the plane surface of this body.
   Use a linear connection !

2. Realize five measurements at least and draw a diagram.

3. Describe your measurement-system in detail.
   Draft it !

4. Give an estimation of the errors.

SOLUTION

(1)

$$T = 2.\pi . \sqrt{\frac{I + I_z}{(M+2m).g.s}} \qquad\qquad I = I_s + M.s^2$$

$$I_z = 2.(I_{cyl} + m.R^2) \qquad\qquad I_{cyl} = \frac{1}{2}.m.r^2$$

$$\longrightarrow\quad T^2 = \frac{4.\pi^2}{(M+2m).g.s} . (I_s + M.s^2 + 2.\frac{m.r^2}{2} + 2mR^2)$$

$$T^2 = \underbrace{\frac{8.\pi^2.m}{(M+2m).g.s}.R^2}_{k} + \underbrace{\frac{4.\pi^2.(I_s + M.s^2 + m.r^2)}{(M+2m).g.s}}_{d}$$

(2)
$$T^2 = k.R^2 + d$$

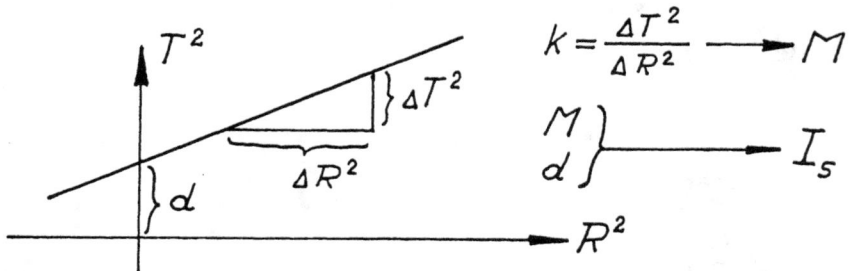

$$k = \frac{\Delta T^2}{\Delta R^2} \longrightarrow M$$

$$\left.\begin{matrix} M \\ d \end{matrix}\right\} \longrightarrow I_s$$

(3)

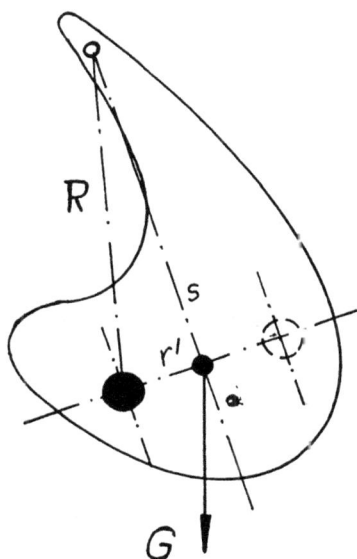

a) Hang up the body and draw centre-of-gravity axes ————► s

b) Determine the centre of mass.

c) Fix the two cylinders symmetrical to the centre of mass. Thereby this point cannot move.

d) $s^2 + r'^2 = R^2$

(4) <u>values</u>

k = 10

d = 0,55

that gives: M = 0,359 kg

$$I_s = 3,55 \text{ kgm}^2 \quad \pm 4\%$$

Günther LECHNER

Theoretical Problem (ÖPHO 1988)

GREETINGS OF A PULSAR ......

SITUATION

The crab-nebula is about 4000 light-years away from us. In its centre there is a star, sending out flashes of lightning. These lightnings hit the earth periodically.

Because of the short time of the lightnings you can find out an interesting detail:
Rays with a wavelength of about 1 m reach us about 1/10 sec later than rays with a wavelength of about 1 cm.

The reason for this is said to be the dispersion of the rays in the interstellare matter.

The matter between the stars mainly consists of homogeniously spreaded protons and electrons.
The density of the interstellar matter ranges between 10^{-24} kg/m³ and 10^{-22} kg/m³.
In this homogeniously spreaded matter there are regions with larger density. They are named "clouds" with densities of about 10^{-20} kg/m³.
These "clouds" absorbe rays with a wavelength of 21 cm.

QUESTIONS

1. Determine which kind of particles is responsible
   for the absorption mentioned above.
   Use a mechanical model !

2. Let us assume a path of the rays <u>not</u> going
   through one of the "clouds":
   Due to the behaviour of the rays of the star
   you are to calculate the density of that part
   of the interstellar matter which you thought
   to be responsible for the dispersion.

   Make comparisons with the result of your cal=
   culation and the given data.
   Do you think this dispersion effect can exist ?
   Give reasons for your answer !

3. Now let us assume a path of the rays going
   through one of the "clouds":

   We assume the border of the "cloud" as a plane.
   Let us imagine that rays of the star hit this
   plane at an angle different to 90° and 0°.

   On the ground of suitable calculations you are
   to draft the paths of the rays before and behind
   the plane border of the "cloud" <u>qualitatively</u>.

HINTS

a. $\Delta v \ll v \Rightarrow \boxed{\dfrac{\Delta v}{v} \approx - \dfrac{L}{v} \cdot \dfrac{\Delta t}{t^2}}$

L ... length of the path

t ... time

b. Let us neglect the damping and the interactions among charged particles in a medium of waves, then the refraction index n can be described with:

$$n = 1 + \frac{N \cdot Q^2}{8\pi^2 \cdot \varepsilon_0 \cdot m \cdot (f_0^2 - f^2)}$$

N .... Number of charged particles in 1 m³

Q .... electric charge of a particle

m .... mass of a charged particle

ε_0 ... permittivity of free space

f_0 ... natural frequency of a charged particle

f .... frequency of the waves

PHYSICAL CONSTANTS

speed of light = $2,9979 \cdot 10^8$ m/s

elementary charge = $1,602 \cdot 10^{-19}$ As

permittivity of free space = $8,85 \cdot 10^{-12}$ F/m

Planck's constant = $6,626 \cdot 10^{-34}$ Js

rest mass of the electron = $9,11 \cdot 10^{-31}$ kg

rest mass of the proton = $1,67 \cdot 10^{-27}$ kg

SOLUTION

1.   density $\longrightarrow$ $N \approx \dfrac{\mathcal{S}}{m(p)}$ $\implies$ $\boxed{600 \longrightarrow 60\ 000\ \dfrac{1}{m^3}}$

Therefore: $N \approx 600 \longrightarrow 60\ 000$ protons <u>and</u> electrons/m³

Distances among the particles are relatively large. You can neglect the interactions between them; these protons and electrons are nearly free.
Hence the natural frequency ≈ 0 Hz.

$$\vec{E} \sim \vec{E_0} \cdot e^{i\omega t} \qquad Q$$

$$P = \frac{2}{3} \cdot \frac{1}{4\pi\varepsilon_0 \cdot c^3} \cdot Q^2 \cdot a^2$$

m

Rays $\longrightarrow$ particles are forced to oszillate:

$\vec{F} = m.\vec{a}$ $W = e.U$ $\boxed{P = \text{prop. } a^2}$

$m(\text{proton}) \approx 2000 \cdot m(\text{electron})$

therefore:

$a(\text{electron}) \approx 2000 \cdot a(\text{proton})$

this gives:

$$\boxed{P(\text{electron}) \approx 2000^2 \cdot P(\text{proton})}$$

If dispersion depends on the interstellar particles, only the <u>electrons</u> can be reason for the dispersion.

2.

$$\frac{\Delta v}{v} \approx - \frac{L}{v} \cdot \frac{\Delta t}{t^2} \implies \frac{\Delta c}{c} \approx - \frac{4 \cdot 10^3 \cdot 365 \cdot 86400 \cdot \frac{1}{10} \cdot 3 \cdot 10^8}{3 \cdot 10^8 \cdot [4 \cdot 10^3 \cdot 365 \cdot 86400]^2}$$

$$\frac{\Delta c}{c} \approx - 8 \cdot 10^{-13}$$

$$\implies \boxed{C_m \approx C_{cm} \cdot [1 - 8 \cdot 10^{-13}]}$$

$$\left. \begin{array}{l} n = 1 + \dfrac{N \cdot e^2}{8 \pi \varepsilon_0 \cdot m \cdot (f_0^2 - f^2)} \\[2mm] f_0 \approx 0 \end{array} \right\} \implies n \approx 1 - \frac{N \cdot e^2}{8 \pi^2 \varepsilon_0 \, m \, f^2}$$

$$n = \frac{C_1}{C_2} = 1 - 8 \cdot 10^{-13}$$

$$1 - 8 \cdot 10^{-13} = \frac{1 - \dfrac{N \cdot e^2}{8 \pi^2 \varepsilon_0 \cdot f_2^2 \cdot m}}{1 - \dfrac{N \cdot e^2}{8 \pi^2 \varepsilon_0 \cdot f_1^2 \cdot m}}$$

$$\implies \boxed{N \approx 1800 \, \frac{1}{m^3}}$$

$$600 < 1800 < 60\,000 \longrightarrow \boxed{\begin{array}{l} DISPERSION \\ POSSIBLE \end{array}}$$

3.

$$g \longrightarrow N \approx \frac{10^{-20}}{1,67 \cdot 10^{-27}} \approx 6 \cdot 10^6 \, \frac{1}{m^3}$$

$$\lambda_0 \approx 21\,cm \longrightarrow f_0 \approx 1,43\ GHz$$

REFRACTION INDEX

before   behind

$\boxed{\lambda = 1\ m}$

$$n_{1V} = 1 - 8 \cdot 10^{-13}$$

$$n_{1H} = 1 + \frac{6 \cdot 10^6 \cdot e^2}{8 \pi^2 \varepsilon_0 m \left[(1,43 \cdot 10^9)^2 - f^2 \right]}$$

$$n_{1H} \approx 1 + 1,2 \cdot 10^{-10}$$

$\boxed{\lambda = 1\ cm}$

$$n_{2V} = 1 - \frac{1800 \cdot e^2}{8 \pi^2 \varepsilon_0 m \cdot (3 \cdot 10^{10})^2}$$

$$n_{2H} = 1 + \frac{6 \cdot 10^6 \cdot e^2}{8 \pi^2 \varepsilon_0 m \cdot \left[f_0^2 - (3 \cdot 10^{10})^2 \right]}$$

$$n_{2V} \approx 1 - 8 \cdot 10^{-17}$$

$$n_{2H} \approx 1 - 3 \cdot 10^{-13}$$

$$n_{1V} < n_{1H} \qquad\qquad\qquad n_{2V} > n_{2H}$$

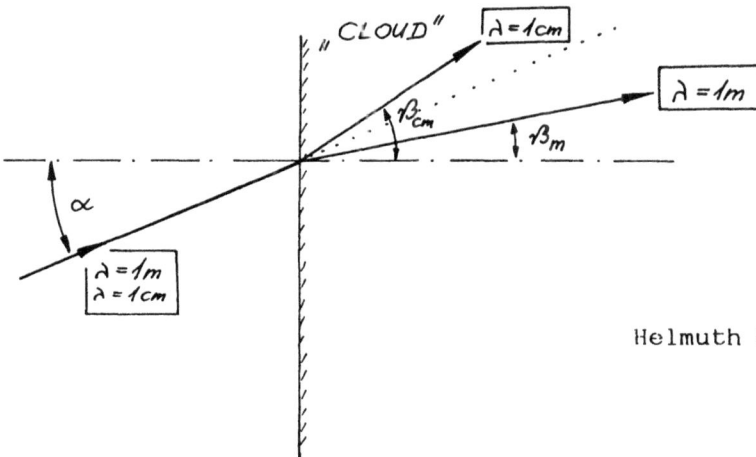

Helmuth MAYR

BULGARIA

BULGARIAN PHYSICS OLYMPIADS

M.H.Maksimov , Sofia University

1. NATIONAL PHYSICS OLYMPIAD.

An annual Physics Olympiad has been held in Bulgaria since 1969. The organization of the Olympiad is in the hands of the National Physics Olympiad Committee in which the Sofia University,the Ministry of Education and the National Society of Physics are represented. The whole organization is financed by the Ministry of Education. The Olympiads are in principle an out-of-school activity. Like the International Physics Olympiad (IPhO) our Olympiad is a competition between individuals. It is organized in four stages.

1.1. First Stage.

The Bulgarian Physics Olympiad starts every year in January, with its first stage, which takes place in schools. About 50000 pupils from the 6th to the 11th school year take part in it. The pupils have to solve two (or three) theoretical problems in two hours. The tests are prepared, checked and evaluated by one of the physics teachers at each school. So, different tests are taken in each school.About 30% of the participants solve the problems successfully and are admitted to the second stage of the Olympiad.

The topics of the first two stages do not overstep the limits of the Physics syllabus in Bulgarian primary and secondary schools. But by the second stage, the difficulty of the tasks to be solved is above ordinary school level.

1.2. Second Stage.

It is called a town stage' and it is held separately for the students of the different grades. Each four-hour test includes three theoretical problems which are prepared by the National Committee. The teachers from the Regional Committees correct and evaluate the papers according to a solution model. The highest possible score is 20 points. The students who score higher than 14 points are admitted to the third stage.

1.3 Third Stage.

Only secondary school students take part in this stage (age of the students between 16 and 18). The topics of the tasks include the full syllabus of physics in secondary school. Most of the students taking part in this stage also take great interest in physics outside school and their knowledge of the subject exceeds the limits of classroom instruction. They have to do a five-hour test. The test is prepared by the National Committee and it includes three theoretical problems. The difficulty of the tasks is above the level of university entrance exams. The papers are checked and marked by the Regional Committees using the solutions and mark scheme provided. The best solutions are then sent to the National Committee,where the 60 most successful candidates are selected to proceed to the final stage.

1.4. Fourth Stage.

The problems of the fourth stage are on the level of the IPhO. This selection lasts for four days and has a theoretical round followed by an experimental one. It is normally organized at the end of April and we do our best to create an atmosphere and conditions similar to the IPhO.

The theoretical test consists of four problems which involve four areas of physics, according to the syllabus of the IPhO.The number of points for each problem is 15. The experimental test includes two laboratory problems.The points available for each problem (theoretical analysis and experimental execution) are 20. The total number of points for both theoretical and experimental rounds is 100.

There is one day of rest between the two rounds.

The competition tasks are chosen , prepared and checked by lecturers from the University of Sofia.

The 10 best winners can enter without futher exams the physics departments of the universities. They are awarded in a special ceremony led by the Minister of Education.

2. THE SYSTEM OF PREPARING THE TEAM FOR THE INTERNATIONAL OLYMPIAD.

An intensive training course is held at the Department of Physics of Sofia University to prepare the students for the IPhO. It is attended by the 10 highest scoring students in stage 4. The training lasts about one month (just before the IPhO) and consists of half-day work sessions which are held by university lecturers according to a plan settled in advance. The program includes theory lectures, problem solving and laboratory work. The purpose of the training is to improve the experimental skills of the students and their theoretical knowledge. About half the time is devoted to theoretical problems and the other half to experiments. During the training the students do several experimental and theoretical tests which are checked and marked.The highest possible score for all tests is 100 points (equal to the score in the 4th stage). The score of these tests together with the score in the 4th stage determine the final order of the students. The five best pupils form the team that will participate in the IPhO. The course usually finishes about a week before leaving for the International Olympiad.

The rest of the pupils, from 9th and 10th school year, who have good results in the 3th and 4th stages, take part in a two-week camp-school in July.Such schools are organized during the winter and spring holidays too.

In several towns in the country specialized out-of-school groups, led by lecturers of physics, work for

about 120 hours yearly. Many of the best Bulgarian
Olympians have been students in such groups.

3. TRAINING PROBLEMS [1, 2]

PROBLEM 1. Let the surface S obtained as a result of
rotation of the curve BB' round the axis X (Fig. 1a)
divide two optical uniform media with indices of
refraction n and n'.

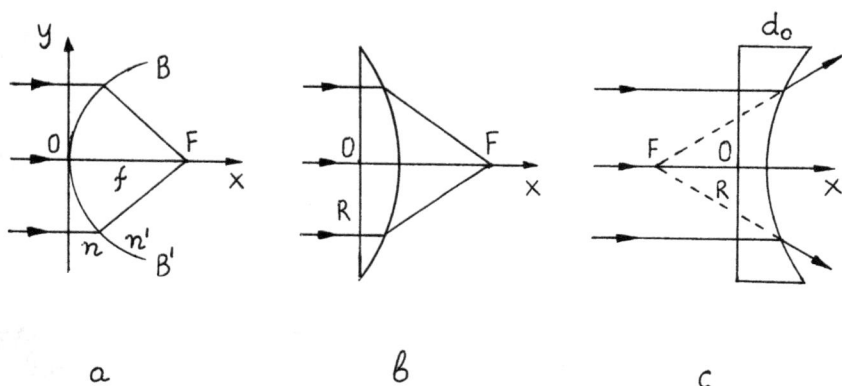

Fig 1.

a) If all rays, parallel to the axis of symmetry X
(the optical axis), after refraction at the surface S
intersect at a single point, which lies on the axis X, S is
called a nonaberration surface.

Obtain the equation for the curve BB' in this case,
if the rays are focused at point F. Given : n, n', OF = f.

Examine the case when n'= −n. Analyse the result.

b) Converging lens with spherical surfaces focuses in
point only paraxial rays. If we want to focus a wide light
beam, we need a lens with nonaberration surfaces.

Determine the minimum thickness in the centre of a
plano-convex lens of refractive index n = 1.5 and radius R
= 5 cm (Fig. 1b) focusing a parallel light beam, which

is incident perpendicularly on the flat face, at point F ;
OF = f = 12 cm.

c) Determine the minimum thickness in the centre of a
plano-concave lens of refractive index n =1.5, radius R = 2
cm and thickness on the periphery d_o= 0.5 cm (Fig 1c) ,
if a parallel light beam falling perpendicularly to the
flat face is refracted in such a way, that the
continuations of the refracted rays intersect at a point F
; OF = f = 20 cm.

PROBLEM 2. At the focus of a paraboloidal mirror of
radius R = 10cm and focal length f = 1m is placed a thin
black disk of sizes coirciding with the image of the Sun at
the focus of the mirror.

What is the highest possible temperature of the disk?

The Sun is a close approximation to a black body of
temperature T_o= 6000K. The thermal conduction of
surrounding air may be neglected

PROBLEM 3. Show that for ordinary (not laser) sources
of light the beam divergence is connected mainly with the
source extension and diffraction divergence can be ignored.

Examine a source of diameter D (for example a
luminous wire) that is placed at the focus of a converging
lens of diameter a and focal length f.

Data: a = 5mm ; f = 5cm; λ =500nm.

PROBLEM 4. Determine the minimum distance δx between
two points that just can be resolved by means of an optical
microscope. The angle subtended by the aperture (objective
lens) is 2θ, the wavelength is λ.

Note: Use Rayleigh's criterion for resolution: two
point sources' diffraction patterns can just be resolved if
the bright centre of one image coincides with the first
dark ring of the other.

Problem 5. Consider a transmittion diffraction grating with a great number of parallel slits having their centres separated by distances d. The width of each following slit is half the width of the previous one. Plane monochromatic light of wavelength λ falls normally on the grating.

Determine the intensity I as a function of observation angle θ for Fraunhofer diffraction through the grating.

Determine the intensity of diffraction maxima and minima.

Problem 6. A weakly divergent beam of singly-charged ions a mixture of ^{39}K and ^{41}K is injected through a narrow aperature into a mass-spectrometer. The kinetic energy of the ions is T = 500±5 eV; the angle of divergence is $2\alpha = 6°$. The uniform magnetic field B = 0.7T is perpendicular to the paper (Fig. 2). A photographic plate is placed on the plane AM.

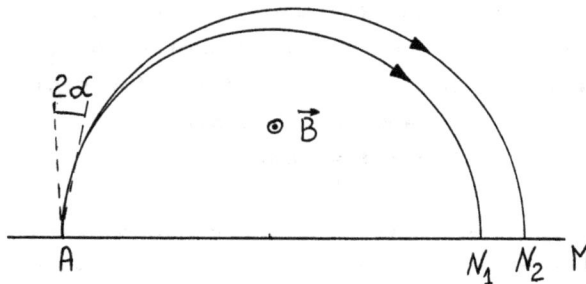

Fig 2.

1. The ions ^{39}K and ^{41}K of energy 500 eV, which are radiated in a straight line towards the plane AM, fall onto points N_1 and N_2 of the plate.

Determine the distances AN_1, AN_2 and N_1N_2.

2. Determine the width of the two isotope lines on

the plate. For that purpose, for one of isotopes calculate:

a) the line broadening Δx_1 caused by the divergence of the beam. Assume all ions have the same energy 500 eV ;

b) the line broadening Δx_2 caused by the energy difference. Assume that all ions are emitted from point A perpendicularly to the plane AM.

The total width of the line is $\Delta x \approx \Delta x_1 + \Delta x_2$.

3. Is it possible to find out the presence of the two isotopes in the beam, i.e. to separate the ^{39}K and ^{41}K lines using this mass-spectrometer?

PROBLEM 7. In metals the free electrons form an ideal gas of indistinguishable particles. That means, the volume, available for each electron to move around in is not the total volume V of the metal, but only a volume V/N , where N is the number of the valent electrons. The average energy E_o which corresponds to one electron includes:

1. The average kinetic energy of thermal motion.

2. The average kinetic energy of "quantum-mechanical" motion. This energy can be evaluated by means of the Heisenberg uncertainty relations ($\Delta x . \Delta p_x \approx \hbar$; .....)

a) Calculate these two energies for sodium (Na) at room temperature $t = 27°C$ and compare them.

b) Calculate the pressure of the electron gas and compare it with atmospheric pressure.

Data: density of sodium $\rho = 9,7.10^2$ kg/m^3, the molar mass of sodium $\mu = 23$ kg/kmol, Planck's constant h $= 6,62.10^{-34}$ J.s, Avogadro's number $N_A = 6,02.10^{23}$ mol^{-1}, the Boltzmann constant k $= 1,38.10^{-23}$ J/K.

c) Evaluate the heat capacity for constant volume, C_V, of the electron gas and compare it with the heat capacity of a classical monatomic gas.

Note: Because of quantum-mechanical effects, the treatment of an electron gas differs considerably from that of an ideal gas. The electron energy E can take only particular discrete values (energy levels). The electrons fill the levels according to the Pauli principle: no more than two electrons can fill one state.

At T=0K the total energy of the electron gas has a minimum— the electrons fill in the first N/2 levels, where N is the number of free electrons. The energy E_F, separating the filled states from the empty states is called the Fermi energy. The approximation $E_F \approx 2E_o$ can be used, where E_o is the average energy of an electron, as determined above.

When the temperature increases, an electron of energy E is included in the heat capacity, i.e. can receive an additional energy $\approx$ kT only if the energy level E + kT is empty (situated above the Fermi level).

PROBLEM 8. a) A relativistic particle, of momentum P , total relativistic energy E and charge q is moving in a circular orbit in a constant uniform magnetic field B.

Determine the cyclotron frequency ω_c and the radius of rotation R.

b) An electron (energy of rest E_o = 0,511 MeV) is moving in a circular orbit of radius R = 10cm in a uniform constant magnetic field B = 0,089T.

Calculate the period of rotation T, the momentum P and the total relativistic energy E of the electron.

PROBLEM 9. The synchrophasotron is a high-energy accelerator where both the transverse magnetic field B(t) and the frequency $\omega(t)$ of the accelerating voltage change simultaneously.

1. Deduce the relation between $\omega(t)$ and B(t) for the motion of the accelerating particles on an orbit of constant radius R.

2. The equilibrium orbit of a synchrophasotron consists of semi-circular and straight sections. On the curved sections, under the action of the transverse magnetic field, the particles move on a circular trajectory

of radius R = 28,0 m, which remains constant during the acceleration. On the straight sections the beam is accelerated and focused. The total path-length of the orbit is L=208m.

Protons (rest energy E_o = 938 MeV) of initial kinetic energy 9,0 MeV are accelerated to the energy of 40000 MeV. During the acceleration the magnetic field inceases at constant rate dB/dt = 0,4 T/s. Determine:

a) the initial and final values of the frequency ν of the accelerating voltage;

b) the total acceleration time Δt;

c) the increase ΔE in the energy of the protons in one revolution;

d) the number of revolutions and total distance travelled by the protons during the acceleration.

The electric field produced by the changing magnetic field is negligible.

PROBLEM 10. a) Deduce that a system of particles of given constant momentum P has a minimum total relativistic energy when all particles are moving in the same diraction at the same velocity.

b) An electron-positron pair (e^- – e^+) can be created by a high energy photon. Fair production was predicted by Dirac (1928). Using the laws of conservation of momentum and energy show that pair production is possible, in this case, only when an additional particle is involved.

c) An electron-positron pair was created in the Coulomb's field of an electron at rest. Determine the minimum (threshold) energy E_t of the photon for this process to be possible.

d) A pair e^- – e^+ was created as a result of the interaction between a photon and an ultrarelativistic electron, moving in the opposite direction to the photon. Find the energy of this electron, if the threshold energy of the photon is E_t = 10eV.

4. SOLUTIONS.

Problem 1. a) We shall assume the rays parallel to the axis X are radiated from point F',which is infinitely far from the point O. The optical path of all rays from F' to F is equal. For a ray incident on the surface S at point A of coordinates x and y the optical path is:

$$S = n \mid F'A \mid + n' \mid AF \mid = const. \qquad (1)$$

But: $\mid F'A \mid = \mid F'A' \mid + \mid A'A \mid$ .All rays cover a distance: $\mid F'A' \mid \approx \mid F'O \mid$, hence:

$$S' = n \mid AA' \mid + n' \mid AF \mid = const. \qquad (2)$$

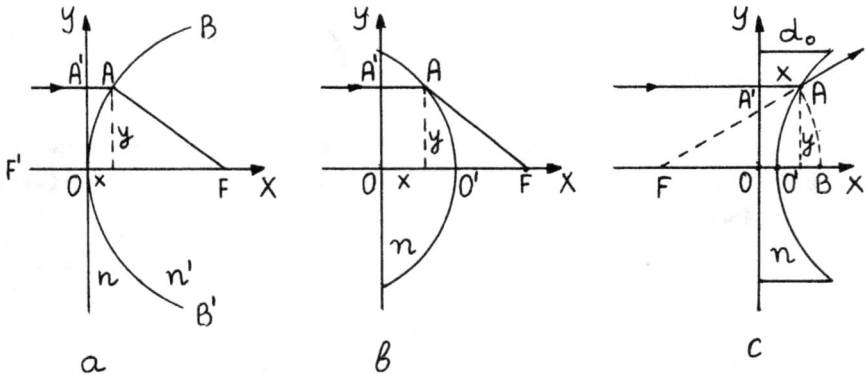

Fig 3.

Using fig.3a we can write:

$$\mid AA' \mid = x \quad ; \quad \mid AF \mid = \sqrt{(f - x)^2 + y^2} . \qquad (3)$$

From (2) and (3) we get:

$$S' = n x + n' \sqrt{(f - x)^2 + y^2} = const. \qquad (4)$$

For the ray coinciding with the axis x :

$$S' = n' \mid OF \mid = n' f . \qquad (5)$$

From (4) and (5):

$$n x + n' \sqrt{(f - x)^2 + y^2} = n' f . \qquad (6)$$

Transfering nx on the right side of equation (6) and squaring we obtain:

$$(n'^2 - n^2) x^2 + n' y^2 - 2 n' (n' - n) f x = 0 . \qquad (7)$$

In general (7) is an equation of ellipse. Then the surface S is an ellipsoid of revolution. Depending on n', n and f, (7) can be an equation of parabola or hyperbola. When n' = -n , from (7) we get: $y^2 = 4fx.$ (8) In this case S is a paraboloidal reflecting surface. Therefore a paraboloidal mirror focuses at a point not only paraxial rays, but a wide light beam as well.

b) Following the method discribed above, from fig.3b we can write:

$$n \mid 00' \mid + \mid 0F \mid - \mid 00' \mid = \mid AA' \mid + \mid AF \mid ;$$

$$\mid 00' \mid = d ;$$

$$n x + \sqrt{(f - x)^2 + y^2} = f + (n - 1) d . \qquad (9)$$

For y = R ; x = 0 from (9) we get:

$$d = \frac{(f^2 - R^2)^{1/2} - f}{n - 1} = 2 \text{ cm.}$$

c) The divergent beam created after refraction at the concave face of the lens we can examine as a beam emitted from an image point source placed at the focus F.

Obviously the plane yOz and the sphere of radius FA and center F are two wave surfaces between which all rays traverse equal optical path (Fig 3c):

$$n \mid A'A \mid = \mid FE \mid - \mid FO' \mid + \mid 00' \mid n ;$$

$$\mid 00' \mid = d ; \mid 0F \mid = f ; \mid AA' \mid = x ;$$

$$n d - (f + d) = n x - \sqrt{(f + x)^2 + y^2} . \qquad (10)$$

For $y = R$; $x = d_0$ and $y = 0$; $x = d$ from (10) we obtain:

$$d = \frac{n\, d_0 + f - (\,(\,d_0 + f\,)^2 + R^2\,)^{1/2}}{n - 1} = 0,3 \text{ cm}$$

Problem 2. According to the Stefan-Boltzmann law, the power Φ_0 radiated by the Sun is:

$$\Phi_0 = 4\pi\, R_\odot^2\, \sigma\, T^4 \; ,$$

where $R_\odot$ is the radius of the Sun. The power Φ_0' emmitted per unit space angle is:

$$\Phi_0' = \frac{\Phi_0}{4\pi} = R_\odot^2\, \sigma\, T^4$$

The power Φ_1 incident upon the mirror is

$$\Phi_1 = \Phi_0'\, \Omega = R_\odot^2\, \sigma\, T^4\, \frac{\pi\, r^2}{L^2} = \sigma\, T^4 \left[\frac{R_\odot}{L}\right]^2 \pi\, r^2 \; ,$$

where: L - the distance from the Sun to the Earth, r - the radius of the mirror. The total light flux, striking the mirror, after reflection falls on the disc. The disk absorbs radiation and its temperature increases. On the other hand the disk radiates power:

$$\Phi_2 = 2\, \sigma\, T_m^4\, S \; ,$$

where T_m is the disk's temperature, S-one side disk's area:

$$S = \pi\, (f\, \vartheta)^2 = \pi\, f^2 \left[\frac{R_\odot}{L}\right]^2 \; ,$$

where ϑ is the angle subtended by the Sun. The temperature is maximum when $\Phi_1 = \Phi_2$. Then:

$$T_m = T_0 \sqrt[4]{\frac{r^2}{2\, f^2}} \approx 1600K \; .$$

Problem 3. The rays, emitted by the edges of the source, after passing through the lens deviate from the optical axis in angle $D/2f$ which determines the geometrical divergence of the beam. The diffraction divergence should predominate if:

$$\frac{D}{2f} < \frac{\lambda}{a} \; ; \quad D < \frac{2\lambda f}{a} = 10 \; \mu m.$$

where λ/a is the angle of diffraction divergence.

The sizes of real sources of light are larger. Increasing the focal length f we could increase admissable sizes of the source, but then the light flux passing through the lens should decrease significantly.

Problem 4. Let's examine a luminous point S_1 placed at the centre of the sight and another point S_2 at a distance δx from S_1. The image of S_1 is a smeared diffraction disk with a bright centre at point S_1' (fig 4).

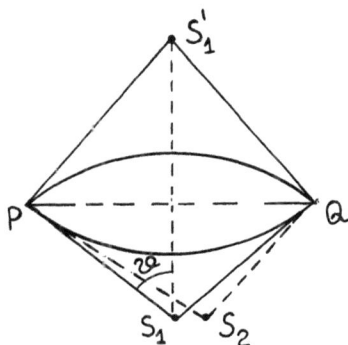

Fig.4.

According to the Rayleigh's criterion, the optical path difference between the rays $|S_2PS_1'|$ and $|S_2QS_1'|$ has to be λ: $|S_2PS_1'| - |S_2QS_1'| = \lambda$.

Let the radius of the objective is r and the distance from the point S_1 to the objective is h. Then:

$$\sqrt{h^2 + (r + \delta x)^2} - \sqrt{h^2 + (r - \delta x)^2} = \lambda \; ;$$

$$\sqrt{h^2+r^2} \left[\left(1 + \frac{2r \; \delta x}{h^2 + r^2} \right)^{1/2} - \left(1 - \frac{2r \; \delta x}{h^2 + r^2} \right)^{1/2} \right] = \lambda \; ;$$

$$2r \; \delta x \; / \; \sqrt{h^2 + r^2} = \lambda \; ,$$

where the terms containing second power of δx are neglected.

$$r \, / \, \sqrt{h^2 + r^2} = \sin \vartheta \qquad ; \; \delta x = \frac{\lambda}{2\sin \vartheta} \quad .$$

Problem 5. In accordance with Huygens-Fresnel's principle each slit may be regarded as a new source of radiation. Its electric-field amplitude E_o is proportional to the width of the slit. So E_o for each following slit is decreased by a factor 2.

For an arbitrary observation point P, far from the grating:

$$E_1 = \frac{E_o}{2} \, e^{i(\omega t - k \, r_1)} \; ;$$

$$E_2 = \frac{E_o}{2} \, e^{i(\omega t - k \, r_2)} \; ;$$

$$\cdot \quad \cdot \quad \cdot \quad \cdot \quad \cdot \quad \cdot \quad \cdot \quad \cdot$$

$$E_n = \frac{E_o}{2^{n-1}} \, e^{i(\omega t - k \, r_n)} \quad .$$

For observation angle ϑ: $\quad r_n = r_1 + (n - 1)d \sin \vartheta \quad .$
The resultant electric field is:

$$E = E_1 + E_2 + \ldots + E_N \; =$$

$$= E_o e^{i(\omega t - k \, r_1)} + \frac{E_o}{2} \, e^{i(\omega t - k \, r_1)} \, e^{-ikd \sin \vartheta} + \ldots =$$

$$= E_o \, e^{i(\omega t - k \, r_1)} \left[1 + \frac{e^{-ikd \sin \vartheta}}{2} + \ldots + \frac{e^{-i(N-1)kd \sin \vartheta}}{2^{N-1}} \right].$$

The sum of the geometric progression in brackets is:

$$S = \frac{1 - 2^{-N} e^{-iNkd \sin \vartheta}}{1 - \frac{1}{2} e^{-ikd \sin \vartheta}} \quad .$$

For a great number of slits N:

$$S = \frac{1}{1 - \frac{1}{2} e^{-ikd \sin \vartheta}} \quad .$$

The electric field is:

$$E = \frac{E_o \, e^{\cdot(\omega t - k \, r_1)}}{1 - \frac{1}{2} e^{-\iota kd \sin \vartheta}} \quad . \tag{1}$$

From (1) we obtain the intensity I as a function of diffraction angle ϑ :

$$I \sim E \, E^{*} = E_o^{2} \Big/ \left[1 - \frac{1}{2} e^{-\iota kd \sin \vartheta} - \frac{1}{2} e^{\iota kd \sin \vartheta} + \frac{1}{4} \right] \; .$$

$$I = \frac{I_o}{1 - \cos (kd \cos \vartheta) + \frac{1}{4}} = \frac{4 \, I_o}{1 + 8 \sin^2(\frac{\pi}{\lambda} d \cos \vartheta)} \quad . \tag{2}$$

From (2) we determine:

$$I_{max} = 4 \, I_o \quad ; \quad I_{min} = \frac{4}{9} I_o \quad ;$$

Problem 6. 1. Due to the magnetic field an ion of mass M moves on a circular path with a radius:

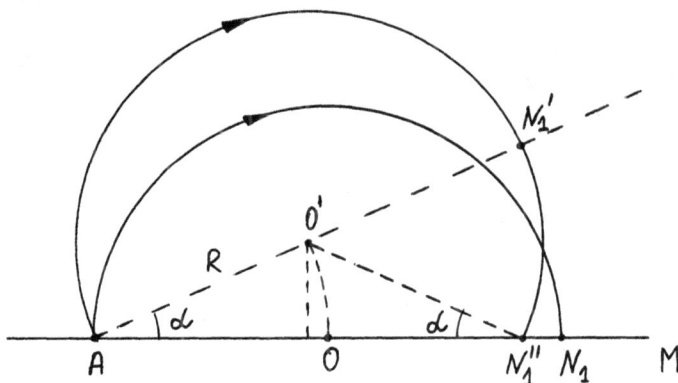

Fig.5.

$$R = \frac{MV}{eB} = \frac{\sqrt{2MT}}{eB} \quad ; \qquad AN = 2R = 2 \frac{\sqrt{2MT}}{eB} \tag{1}$$

For ^{39}K : $AN_1 = 5,75$ cm.

For ^{41}K : $AN_2 = 5,90$ cm ; $N_1 N_2 = 1,45$ mm

2a) The ions entering perpendicularly to the plane AM

move on a semicircle AN_1 with a radius R. The ions , emitted in direction α towards the normal of the plane AM move on the catenery $AN_1'N_1$ with the same radius R.

Using the plan of fig. 5 we can write:

$$N_1N_1'' = AN_1 - AN_1'' = 2R - 2R \cos \alpha =$$

$$= 2R(1 - \cos \alpha) = 4R \sin^2 \frac{\alpha}{2} \approx R\alpha^2.$$

The ions, emitted symmetrically (in direction $-\alpha$), fall into the same point N_1''.

$$\Delta x_1 = R \alpha^2 = 0,08 \text{ mm} \tag{2}$$

2b) Substituting $T = T_0 + \delta T$ in (1) we get:

$$d = 2 \frac{\sqrt{2MT_0}}{eB} \left[1 + \frac{\delta T}{T_0} \right]^{1/2} \approx 2R \left[1 + \frac{1}{2} \frac{\delta T}{T_0} \right]$$

$$\Delta x_2 = d_{max} - d_{min} = 2R \frac{\Delta T}{T_0} = 0,58 \text{ mm} \tag{3}$$

$$\Delta x = \Delta x_1 + \Delta x_2 = 0,66 \text{ mm}$$

3. $N_1N_2 \approx 2\Delta x$. The line are clearly separated.

Problem 7. We can assume that each electron is located in a cube of volume V/N and edge $a = \sqrt[3]{\frac{V}{N}}$. Then from the uncertainty relation we get:

$$P_x a = \hbar \quad ; \quad P_y a = \hbar \quad ; \quad P_z a = \hbar$$

$$P = \sqrt{P_x^2 + P_y^2 + P_z^2} = \sqrt{3} \frac{\hbar}{a}$$

The average kinetic energy of quantum-mechanical motion is:

$$\overline{T}_0 = \frac{P^2}{2m} = \frac{3}{2} \frac{\hbar^2}{m} \frac{1}{a^2}$$

On the other hand using that the number N of valent electrons is equal to the number of atoms, we obtain:

$$\frac{V}{N} = \frac{V}{M} \frac{M}{N} = \frac{1}{\rho} \frac{\nu \mu}{\nu N_A} = \frac{\mu}{\rho N_A} \cdot$$

$$\overline{T}_0 = \frac{3}{2} \frac{\hbar^2}{m} \left[\frac{N_A \rho}{\mu} \right]^{2/3} \approx 1 \text{ eV}$$

The average thermal energy is:

$$\overline{T} = \frac{3}{2} kT \approx 0,04 \text{ eV} \ll \overline{T}_0$$

b) The pressure of the electron gas is:

$$p = \frac{2}{3} n \bar{T}_0 = \frac{N_A \rho}{\mu} \frac{\hbar^2}{m} \left(\frac{N_A \rho}{\mu} \right)^{2/3} = \frac{\hbar^2}{m} \left(\frac{N_A \rho}{\mu} \right)^{5/3}$$

$$p = 2,7.10^9 \ Nm^{-2} = 2,6.10^4 \ atm$$

c) At temperature T we can consider only electrons of energy from $E_F - kT$ to E_F to take part in the thermal motion. The electrons of lower energy E can not be thermally activated, since for them the levels of energy E + kT are occupied. We shall assume a uniform distribution of the states (constant density of states).Then $\Delta N = N \frac{kT}{E_F}$ electrons are located in the range $(E_F - kT ; E_F)$.

The total thermal energy E_T of the electron gas is:

$$E_T = \frac{3}{2} kT\Delta N = \frac{3}{2} \frac{k^2 T^2}{E_F} N$$

The heat capacity is:

$$C_V = \frac{dE_T}{dT} = \frac{3}{2} \frac{2k^2 T}{E_F} N$$

For ideal monatomic gas:

$$C_V^{id} = \frac{3}{2} k N$$

Hence:

$$C_V = C_V^{id} \frac{2kT}{E_F} \approx C_V^{id} \frac{k^-}{E_0} \ ; \ at \ T = 300 \ K \ \ C_V \approx 0,026 \ C_V^{id}$$

Therefore only a small part of the free electrons can be thermally activated and takes part in the electron heat capacity. The number of these electrons increases linearly with raising the temperature (at not too high temperatures when $kT \ll E_F$). The electron heat capacity has the same linear temperature dependence.

Problem 8.1. The Lorentz force F_L = qvB acting on a charge q is perpendicular to the velocity (momentum). The force F_L changes only the direction of the momentum of the particle and does not change its magnitude. Rotation $d\varphi$ in time dt is:

$$d\varphi = \frac{dP}{P} \qquad (1) \qquad (fig \ 6).$$

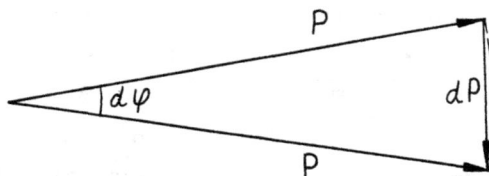

Fig.6.

We can apply Newton's second law:

$$\frac{dP}{dt} = qVB \tag{2}$$

Hence:

$$\omega_c = \frac{d\varphi}{dt} = \frac{qVB}{P} \tag{3}$$

Using the relativistic relation $V = Pc^2/E$ we get:

$$\omega_c = \frac{qB}{E/c^2} \tag{4}$$

On the other hand $V = \omega_c R$ and from (3) we obtain the radius R of the orbit:

$$R = \frac{P}{qB} \tag{5}$$

2.   2,1 ns ; 2,67 MeV/c ; 2,2 MeV

Problem 9. 1. The momentum P and cyclotron frequency ω_c of the particle of charge q, moving in a transverse magnetic field B, are:

$$P = qBr$$

$$\omega_c = \frac{qBc^2}{E} = \frac{qBc^2}{\sqrt{P^2c^2 + m_o^2 c^4}}$$

Using the condition for acceleration $\omega_c = \omega(t)$ we get:

$$\omega(t) = \frac{c}{r} \frac{1}{\sqrt{1 + \left(\frac{m_o c}{qrB(t)}\right)^2}} .$$

2a). The frequency of the accelerating voltage is:

$\nu = \frac{1}{\tau}$;where τ is the time per one revolition.

$$\tau = \frac{L}{V} = L \ / \ \left(\frac{Pc^2}{E} \right) \quad ;$$

$$\nu = \frac{Pc^2}{EL} = \frac{c^2\sqrt{E^2 - E_0^2}}{ELc} = \frac{c}{L} \sqrt{1 - \left(\frac{E_0}{E} \right)^2} \ , \tag{1}$$

where the increase of total energy E during one revolution is neglected.

$$\nu_{in} = 0,20 \ \text{MHz} \quad ; \quad \nu_{fin} = 1,44 \ \text{MHz}$$

2b). For the shaped section:

$$\omega = \frac{qBc^2}{E} \tag{2} \quad ; \quad P = qBr \tag{3}$$

From (3) we get:

$$P - P_0 = qr(B - B_0) \quad ; \quad \frac{P - P_0}{qr} = \Delta B \ . \tag{4}$$

On the other hand:

$\Delta B = \frac{dB}{dt} \Delta t$,where Δt is the total time of acceleration, ΔB – increase of B in time Δt, P_0 and P are initial and final values of the momentum:

$$\Delta t = \frac{P - P_0}{qr \ \frac{dB}{dt}} = 3,2 \ \text{s}$$

2c). $\Delta E = \bar{F} L$ (5),where $\bar{F}$ is an average force which is directed tangentially to the trajectory. The force $\bar{F}$ changes only the magnitude of the momentum:

$$\bar{F} = \frac{dP}{dt}$$

Using the condition $r = const$ from (3) we get:

$$\frac{dP}{dt} = qr \ \frac{dB}{dt} \ . \tag{6}$$

From (5) and (6) we obtain:

$$\Delta E = qrL \ \frac{dB}{dt} = 2,33 \ \text{keV}.$$

2d).$N \ \Delta E = E - E_0$; N – number of revolutions, $E - E_0$ –total increase of proton energy during the acceleration.

$$N = \frac{E - E_0}{\Delta E} = 4,3 . 10^6 \ \text{revolutions} \ ; \quad S = N \ L = 9 . 10^5 \ \text{km}.$$

Problem 10.1. The total energy E and the total momentum P of an isolated system of particles depend on the frame of reference where they are measured. However the quantity $E^2 - P^2c^2$ is the same in all inertial frames of reference and it is called relativistic invariant. In the centre-of-mass frame $P_c = 0$. Hence:

$$E = P^2c^2 + E_c^2 \qquad (1)$$

At given momentum P in the laboratory frame (l-frame), the energy E has a minimum value when the total energy E_c in the centre-of-mass frame (c-frame) is minimum.

$$E_c = \sum m_{oi}c^2 + \sum T_{ci} \ ,$$

where m_{oi} and T_{ci} are the mass of rest and the kinetic energy of i-th particle. Since T_{ci} is always positive (or zero), E_c is minimum when $\sum T_{ci} = 0$, i.e. when in the l-frame all particles are moving in a straight line with same velocity, equal to the velocity of the centre-of-mass.

3. The energy of photon will be minimum when after reaction the three particles move in a diraction of the falling photon with the same momentum P (see 10.1).

From the laws of conservation of momentum and energy it follows that:

$$P_\gamma = 3P \ , \qquad (1)$$

where P_γ is the momentum of the photon.

$$E_\gamma + E_o = 3E \ , \qquad (2)$$

where E is the total energy of each particle, E_o – electron rest energy.

From the relativistic invariant:

$$P = \frac{1}{c}\sqrt{E^2 - E_o^2} \quad (3) \qquad ; \qquad P_\gamma = \frac{E_\gamma}{c} \ . \qquad (4)$$

Using the equation (2), substituting (3) and (4) into (1) and squaring we get:

$$E_\gamma = 4E_o. \qquad (5)$$

For electron $E_o = 0,511$ MeV and the threshold energy is $E_t = E_\gamma \approx 2$ MeV.

4. In the l-frame the electron moves with a velocity v close to the velocity of light. We shall examine the problem in the dynamical reference frame, where the electron is at rest. Using the Lorentz transformations for energy and momentum we obtain the energy of photon in this frame:

$$E' = \frac{E + P.V}{\sqrt{1 - V^2/c^2}} \quad , \tag{6}$$

where E and P are the energy and the momentum of the photon in the l-frame. The sign "+" in the numerator of (6) shows that the photon is moving against the electron. Substituting $\frac{V}{c} = \beta$; $P = \frac{E}{c}$ and $E = E_t$ into (6) we get:

$$E' = E_t \sqrt{\frac{1 + \beta}{1 - \beta}} \quad . \tag{7}$$

Since $\beta \approx 1$ (ultrarelativistic electron), then:

$$E' \approx \frac{E_t \sqrt{2}}{\sqrt{1 - \beta}} \tag{8}$$

The total energy E_e of the electron is:

$$E_e = \frac{E_o}{\sqrt{1 - \beta^2}} = \frac{E_o}{\sqrt{(1-\beta)(1+\beta)}} \approx \frac{E_o}{\sqrt{2(1-\beta)}} \tag{9}$$

Divading equation (9) by equation (8) gives:

$$E_e = \frac{E' . E_o}{2 . E_t} \tag{10}$$

In our case :

E'= 2 MeV (the threshold energy in the dynamical frame of reference); E_o= 0,5 Mev ; E_t = 10 eV.

From (10) we get: E_e = 52 Gev .

5.REFERENCES.

1.M. Maksimov and B. Karadjov, Problems in Atomic and Nuclear Physics (MCSE, Sofia, 1988).

2.M. Maksimov , B. Karadjov and S. Stoyanov, Problems in Optics (MCSE, Sofia, 1988).

CANADA

CANADIAN PHYSICS OLYMPIAD

John Wylie
The Toronto French School
306 Lawrence Avenue E.,
Toronto, Ontario
Canada   M4N 1T7

Michael J. Crooks
Department of Physics
University of British Columbia
Vancouver, British Columbia
Canada   V6T 2A6

Introduction

Until quite recently, the International Physics Olympiad seems to have been almost unknown amongst Canadian physicists. Yet this demanding and prestigious academic competition for high school students has been held (almost) annually since it originated in Warsaw in 1967. After an initial period when the competitors were entirely from eastern Europe, the scope of the Olympiad has broadened, and is now truly world wide.

Each country is represented by a team of five students from general or technical secondary schools, who are not yet 20 years old, and the team is accompanied by two leaders. In 1985 the first team of five Canadian students attended the competition, which was held in Portoroz, Yugoslavia, and there has been a Canadian team at each succeeding Olympiad.

The Olympiad competition has both theoretical and experimental components. The syllabus for the theoretical exam covers most of the topics which might be seen in a fairly intensive first year university physics course. The typical level of achievement is that of a good first (or even second) year university student. However, the extensive use of calculus is not expected. Such a range of material may seem too demanding for Canadian students, but from our experience, the students who make up the teams often have problem solving skills which are as good as, or better than, those of many second-year physics majors at Canadian universities.

The practical tasks set to the competitors involve the design, construction, execution and analysis of experiments using equipment which, in the past, has varied from common household items to sophisticated optical and electronic devices. Students are expected to have a grasp of the rudiments of experimental errors. Clearly, the students who do well have exceptional motivation and ability, and are prepared well beyond what is usual in secondary school. Medal winning Olympiad competitors have demonstrated their ability to apply their knowledge in the laboratory.

The Canadian Programme

All expenses for the duration of the Olympiad are borne by the local hosts, and this extends to the provision of spending money for each of the participants. The matter of selecting a team and getting to the Olympiad is the responsibility of each of the national organizations and the details obviously vary a great deal from country to country.

It is our policy that no Canadian student trying for the team, should face any personal expenses. Students must take part in training programmes offered by a participating university in their province or with the national organization. Every school in Canada is sent information on how students may become involved and, this year, we estimate that over 500 students took part in the selection process. All programmes culminate with the writing of the national selection exam in May. Across Canada, the degree of participation in the various training programmes unfortunately varies a great deal from region to region. The five teams which have thus far attended the International Olympiads have come from seven of the ten provinces, with the heaviest representations from Ontario, Quebec and British Columbia, probably because of the more intensive programmes in those provinces.

The Olympiads operate in both of our National Languages— English and French. A Francophone student has yet to be selected to the Physics Olympiad Team but such is not the case for the Chemistry Olympiad in Canada.

The University of British Columbia's programme is typical of what we are trying to achieve with our training. All grade 11 and 12 physics teachers in the province are written in September asking for the nominations of their brightest prospects. Monthly problem sets are then sent out to these students and also to any teachers who have expressed interest. The students submit their worked problems to the university where they are marked and returned. Based on the results of these problems and also the results on the provincial CAP—Canadian Association of Physicists— high school competition, a group of 20 to 25 students are invited to the university for a three day stay in May. They spend the time in laboratory work, seminars and social events and finish their stay by writing the national selection exam.

Such provincial programmes are not yet set up in all areas of Canada. A student who cannot participate in a provincial programme because of location is sent preparatory materials in the fall and is sent a screen test in February. This test is used as the basis for inviting students to write the May selection Exam.

Results at the Olympiads have been satisfactory, but a Canadian student has yet to place above the bronze medal level. In part, that is certainly due to the lack of an intensive study program with the team. In 1989, we succeeded in raising sufficient funds to remedy this deficiency. Shortly after the selection exam, a select group of twelve students was invited to a central Olympiad training camp at the University of British Columbia.

The camp involved four days of intensive training plus two travel days which were combined with social activities. A typical day involved a morning lecture followed by a problem session and a short test. In the afternoon a full laboratory problem was set to the students and in the evenings, the students had practice problems based on the day's material to work on. Among the non-academic highlights of the students' experience were a tour of Vancouver, a visit to the TRIUMF physics research facility and a formal banquet hosted by the University President's office.

At the end of their stay, the Canadian Team was announced to the students at a special breakfast ceremony. In addition to fielding a stronger team, the training camp also enables us to stimulate and reward a larger number of bright young Canadian physicists.

Olympiad Organization

The Canadian organization and team selection for the International Olympiad is organized by the Canadian Chemistry and Physics Olympiad, a charitable, non-profit corporation. A Board of Directors meets two or three times a year to plan Olympiad policy, fundraising and direction. The Board appoints an Executive Officer who oversees the day-to-day organization and fundraising. A great many people work voluntarily with the Olympiads as Academic Committee Members, Provincial Programme Directors, Training Camp Organizers, and Team Leaders.

Canada has no Federal Ministry of Education nor any government body whose mandate would include organizing Olympiads. Our Chemistry and Physics Olympiads are organized under private initiative. Our annual operating expenditures for both Olympiads is now some $100,000 (Canadian). It is our obligation and pleasure to recognize the generous support we have received.

The Canadian Chemistry and Physics Olympiads were founded by The Toronto French School. The sponsors of the Canadian Olympiads are:

Toronto French School   National Research Council

Recognition should be extended to the following for their financial support of the Olympiads.

Bell Northern Research	Dow Chemical
Investors Group Inc.	Domtar Chemicals
Tioxide Canada	Shell Canada
Xerox Research	John Wiley & Sons Canada
Glaxo Canada Inc.	Du Pont Canada
The McLean Foundation	The Boland Foundation
Ciba-Geigy Canada	Province of Quebec

As well, the following Ministries of Education have sponsored—or have a policy for sponsoring—team members from their provinces to the International competitions.

British Columbia	Saskatchewan
Northwest Territories	Yukon
New Brunswick	Nova Scotia
Alberta	Ontario

The Universities of British Columbia, Toronto and Bishop's and Dalhousie University support the Olympiads through their extensive training and selection of Canadian students.

Benefits

The benefits of participation in the Olympiad competitions exist at many levels, and after several years, they are still developing. Clearly, the pleasures and satisfaction of representing their country at an international competition are a fine reward for demonstrated excellence for the five students who make up the team. The event offers them a unique chance to meet their peers from around the world, and many of the leaders know of enduring friendships which have developed from meetings at the Olympiads.

The Canadian students clearly feel that they represent their country, and though we, as leaders, do not feel in any sense that "national prestige" is associated with the team's achievements, the students do take pride in demonstrating some distinction.

More important benefits accrue to the much larger number of students who take part in the preparation programmes. These are all excellent students, typically only one or two at a school, and they are given a chance to sharpen their skills by tackling a number of problems at a level they are not likely to encounter in their texts or in their class rooms. The students enjoy the challenge, and secondary school teachers have frequently commented to the organizers how helpful they find the tutorial problems in stimulating the students who get involved. Many teachers like to receive the problem sets even when they do not have students working on them, so the occasional problem, or some modification, may find its way into a few classrooms. The programme can, in this way, hope to provide assistance and stimulation to the teachers.

In addition, a number of Canadian universities offer entrance scholarships to any student who makes either of the Physics or Chemistry Olympiad teams and who meets with their usual entrance requirements.

But probably the most important benefit is that enthusiastic physics students, while still at secondary school, whether or not they participate, can come to see themselves as part of a larger group than just the local physics class. Here is a programme, world-wide in scope, which they can take part in, and which recognizes as important just the kind of talents they possess. By exposing our most promising science students to demanding theoretical and experimental work of an international standard, we hope to make an impact on the quality of physics education in Canada and to excite students about the realistic possibilities of physics as a career.

1989 Canadian Selection Exam

This exam was used in the second stage of our National Competition. Students with a mark of 45% or more were invited to the National Training Camp. The top mark was 75%. The school curricula in Eastern (and many parts of Western) Europe are very different from those generally encountered in North America. It is unlikely that even the brightest Canadian student would fare well on the Olympiad without a considerable amount of preparation. The exam questions are meant to stretch the best students— those who <u>have</u> studied their curriculum intensively.

Question 1

You have probably been taught that a ball thrown upwards into the air will take as long to reach its maximum altitude as it will to fall back to its original position. Is this true? In no more than 5-6 lines, justify your answer.

Question 2

An airplane flies a straight course from A to B and back again. The distance between A and B is L and the airplane maintains a constant airspeed V. There is a steady wind with a speed v.

(a) Calculate the total time for the trip if the wind blows along the line AB. Ignore any turn around times.

(b) Calculate the total time for the trip if the wind blows perpendicularly to the line AB.

(c) Find an expression for the total trip time for an arbitrary wind direction. Notice that the trip time is always increased by the presence of a wind in any direction.

Question 3

A beam of electrons from passes through a region of space containing a uniform electric field $\vec{E}$ and a uniform magnetic field $\vec{B}$ (Fig. 1).

(a) If the electrons have a velocity v_0, what conditions must be satisfied by the electric and magnetic fields for there to be no deflection of the particles?

(b) Now assume that you turn off the magnetic field, that the length of the region containing the electric field is L, that the electric field is perpendicular to the initial velocity, and that the distance to a florescent screen is D. What accelerating voltage V is required to produce the deflection y shown.

Question 4

A pendulum (Fig. 2) is constructed from a rigid rod of length L and negligible mass. The pendulum bob has mass M and is fastened to the wall with a horizontal spring of force constant k. When the pendulum hangs straight down the spring is relaxed as shown. Find the period of oscillation of the system for "small" values of the amplitude.

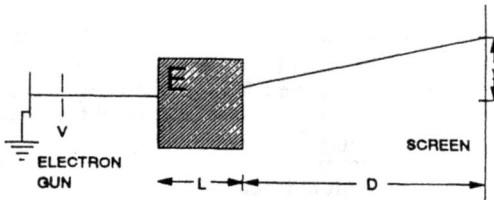

Fig. 1

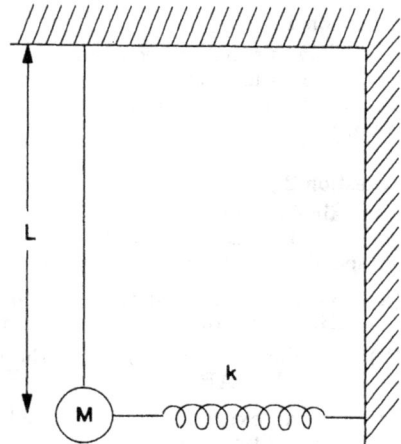

Fig. 2

Question 5

A power supply of constant voltage V_1 is connected to a transformer with N_1 and N_2 primary and secondary windings (Fig. 3). The windings have an internal resistance of r_1 and r_2 respectively. The unit is used to power a load of variable resistance R.

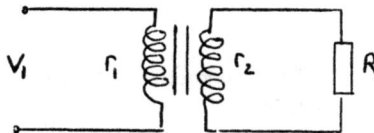

Fig. 3

Find a relation between the power P_1 drawn from the power supply and the power P_2 dissipated by the load R. Make a sketch of the graph of P_1 versus P_2 (i.e. as R is varied).

Question 6

A band was marching on a one-way street in the same direction as the normal traffic flow. A motorist was approaching in his car from behind the band just as they played, in unison, a 440 Hz musical note. The same band's performance was being broadcast "live" via a stationary microphone on the street just ahead of the band. The motorist was listening to the broadcast on the car radio.

The motorist noticed that the combination of the direct sound from the band ahead of him and from the radio produced discernible beats. A measurement counted four beats in three seconds. From the speedometer, the motorist noted the car's speed as 18 km/hr. Calculate the speed of the marching band given that, on this cold day, the speed of sound was 330 m/s.

Question 7

A transparent film of thickness 5.2×10^{-5} cm is viewed under white light from an angle of 31° to the normal. The refractive index of the film is 1.35. Determine the wavelengths of light in the visible spectrum (380–780 nm) which will be absent from the reflected light. Predict the appearance of the film under reflected light.

Question 8

The two cylinders (Fig. 4) shown have cross-sectional areas 10^{-2} m². The left-hand cylinder contains 4 g of a gas at 0°C and at 1 atm (1.013×10^5 Pa) pressure. Its volume is 22.4 l. The right-hand cylinder contains 7.44 g of the same gas, also at 0°C and of volume 22.4 l. The walls of the left-hand cylinder are thermally insulated whereas the right-hand cylinder is maintained at 0°C by a reservoir. The whole system is in vacuum.

When released, the rigidly connected pistons move 0.5 m to their equilibrium point. The gas has a specific heat capacity (at constant volume) of $c_v = 3.14 \times 10^3$ J Kg^{-1} K^{-1}. How much heat did the gas in the right hand cylinder take up and in what form? What is the specific heat, at constant pressure, of the gas?

Fig. 4

Solutions

Question 1

Air resistance has the effect of reducing the maximum altitude of the ball but, more significantly, as the ball falls it approaches (but perhaps does not reach) a terminal velocity. Thus the ball takes longer to fall back than to rise. In an extreme case, imagine what happens when a feather is thrown upwards.

Question 2

(a) For a wind blowing along the line AB we have for the total trip time,

$$T = \frac{L}{V+v} + \frac{L}{V-v}$$
$$= \frac{2VL}{V^2 - v^2}$$

(b) For a wind blowing perpendicularly to the line AB, the wind speed and the plane's airspeed must add like vectors to give the plane's ground speed. This speed is the same for both legs of the flight so that

$$T = \frac{2L}{\sqrt{V^2 - v^2}}$$

(c) For the general case we specify the wind direction by the angle θ from the line AB. The direction of the plane's airspeed is specified by an angle α similarly defined. The plane's ground speed has magnitude v_1 and is directed along the line AB. Hence

$$v \cos\theta + V \cos\alpha = v_1 ,$$
$$v \sin\theta + V \sin\alpha = 0 .$$

The time for the first leg of the trip is, upon solving for v_1, given by

$$t_1 = \frac{L}{v_1} ,$$
$$= \frac{L}{V\sqrt{1 - \frac{v^2}{V^2}\sin^2\theta} + v\cos\theta} .$$

For the return leg we have

$$v \cos\theta + V \cos\alpha = -v_2 ,$$
$$v \sin\theta + V \sin\alpha = 0 ,$$

where we note that $\cos\alpha$ must now be negative since the plane is travelling in the opposite direction. The time for the return leg is

$$t_2 = \frac{L}{v_2} ,$$
$$= \frac{L}{V\sqrt{1 - \frac{v^2}{V^2}\sin^2\theta} - v\cos\theta} .$$

The total time is finally

$$T = t_1 + t_2 ,$$
$$= \frac{2VL}{V^2 - v^2}\sqrt{1 - \frac{v^2}{V^2}\sin^2\theta} .$$

Question 3

(a) For no deflection, the magnitudes of the fields must obey $v_0 = E/B$. The directions of the fields must be in the configuration shown.

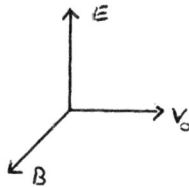

(b) The deflection y is given by

$$y = s + D \tan \theta \ ,$$

where s is the vertical deflection of the electron beam immediately upon leaving the electric field and θ is the angle of the deflection of the beam (above the horizontal). The vertical acceleration of the electrons while in the electric field is eE/m so that basic kinematics yields

$$s = \frac{eEt^2}{2m} \ ,$$

$$\tan \theta = \frac{v_y}{v_x} \ ,$$

$$v_x = \frac{L}{t} \ ,$$

$$v_y = \frac{eEt}{m} \ .$$

Clearly $v_x = v_o$ and the accelerating voltage gives each electron an amount of kinetic energy given by

$$eV = \frac{1}{2}mv_o^2 \ .$$

Putting this all together gives the required result,

$$V = \frac{EL}{2y}\left(\frac{L}{2} + D\right) \ .$$

Question 4

We assume that the rod is long so that the moment of inertia of the pendulum about its pivot is just ML^2. We further assume that the spring is long such that for small amplitude oscillations, the spring is essentially horizontal. We construct the force diagram

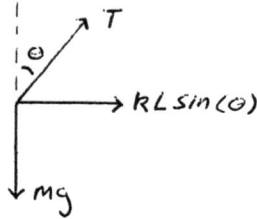

which, upon resolving components, yields the torque equation

$$-(mg\sin\theta + kL\sin\theta\cos\theta)L = ML^2\ddot{\theta} \ .$$

For small angles this simplifies to give the equation of motion

$$\ddot{\theta} + \left(\frac{g}{L} + \frac{k}{m}\right)\theta = 0 \ .$$

The coefficient of the θ term is ω^2 where ω, the angular frequency, is related to the period of oscillation by $\omega = 2\pi/T$. The period is then

$$T = 2\pi\sqrt{\frac{mL}{mg + kL}} \ .$$

Question 5

We know that for the transformer circuit given,

$$I_1V_1 = I_2V_2 \ ,$$
$$V_2 = I_2(r_2 + R) \ ,$$
$$N_2V_1 = N_1V_2 \ ,$$

where V_2 is the voltage induced across the secondary windings. The powers are given by $P_1 = I_1V_1$ and $P_2 = I_2^2R$. It is straightforward to show that

$$P_1 = P_2 + P_1^2 \cdot \frac{N_1^2 r_2}{N_2^2 V_1^2} \ .$$

This is the equation of a parabola. If we define $a = N_2^2 V_1^2/N_1^2 r_2$, then we have

$$\left(y - \frac{a}{2}\right)^2 = -a\left(x - \frac{a}{4}\right) \ .$$

The graph is of the form

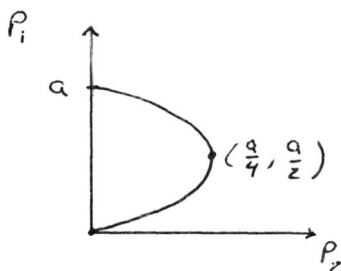

Question 6

From the doppler shift for sound, the frequency of the sound heard directly is

$$f = f_0 \left(\frac{1 + v_c/c}{1 + v_b/c} \right) ,$$

where f_0 is the frequency produced by the band, c is the speed of sound in air, v_c is the car's speed and v_b is the speed of the band. The frequency heard on the radio (remember that the microphone is in front of the marching band) is

$$f' = \frac{f_0}{1 - v_b/c} .$$

Given that the beat frequency is

$$f' - f = \frac{4}{3} \text{ Hz} ,$$

it is straightforward to solve for the speed of the marching band,

$$v_b = 2.98 \text{ m/s} .$$

Question 7

From the ray diagram shown above, the effective path difference for the two rays is

$$\delta r = \frac{2dn}{\cos r} - s + \frac{\lambda}{2} ,$$

where d is the thickness of the film, n is the index of refraction of the film and λ is the wavelength of the light under consideration.

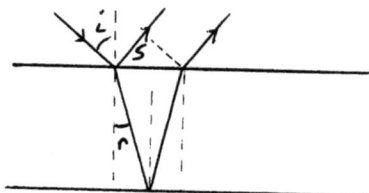

From Snell's law, $\sin i = n \sin r$, we find that

$$s = \frac{2d \sin^2 i}{\sqrt{n^2 - \sin^2 i}} \ .$$

The condition for destructive interference (we are looking for missing wavelengths) is

$$\delta r = (2m + 1)\frac{\lambda}{2} \ .$$

We find, after a little algebra, that missing wavelengths satisfy

$$\lambda = \frac{2d\sqrt{n^2 - \sin^2 i}}{m} \ , \ m = 1, 2, 3 \ldots \ .$$

For $m = 2, 3$ we find two wavelengths in the visible region which are missing. They are 649 nm (red) and 433 nm (violet). The reflected light likely appears yellow-greenish.

Question 8

The pressure of the right-hand gas, P_r, is given by

$$\frac{P_l}{P_r} = \frac{n_l}{n_r} \ ,$$

where P_l is the pressure of the left hand gas, and n_l and n_r are the number of moles of the left- and right-hand gases respectively. Clearly

$$P_r = 1.013 \times 10^5 \left(\frac{7.44}{4}\right) \text{ Pa} \ ,$$
$$= 1.884 \times 10^5 \text{ Pa} \ .$$

At equilibrium, the left- and right-hand gases have volumes V_l and V_r respectively.

$$V_l = (22.4 \times 10^{-3} - 0.5 \times 10^{-2}) \text{ m}^3$$
$$= 17.4 \times 10^{-3} \text{ m}^3 \ ,$$

$$V_r = (22.4 \times 10^{-3} + 0.5 \times 10^{-2}) \text{m}^3$$
$$= 27.4 \times 10^{-3} \text{m}^3 \ .$$

At equilibrium, the pressures P in the two cylinders are equal. The gas in the right-hand cylinder expands isothermally (PV is constant) so that

$$P = 1.884 \times 10^5 \left(\frac{22.4}{27.4}\right) \text{ Pa}$$
$$= 1.540 \times 10^5 \text{ Pa} .$$

The left-hand gas is compressed adiabatically (PV^γ is constant where $\gamma = c_p/c_v$). Hence

$$\frac{1.013}{1.540} = \left(\frac{17.4}{22.4}\right)^\gamma ,$$
$$\gamma = 1.66 .$$

The final temperature of the left-hand gas is from the equation of state of an ideal gas (PV/T is constant).

$$T = \frac{1.54 \times 17.4}{1.013 \times 22.4} \times 273 \text{ K} ,$$
$$= 322.4 \text{ K} .$$

The right-hand gas takes up the heat given off by the left-hand gas (conservation of energy) in the form of internal energy. Remember that the right-hand gas's temperature cannot rise. This heat is calculated by finding the loss of heat from the left-hand gas.

$$Q = c_v m \delta T ,$$
$$= (3.14 \times 10^3)(4 \times 10^{-3})(322.4 - 273) \text{ J} ,$$
$$= 620.5 \text{ J} .$$

Finally, the constant pressure specific heat is simply given by

$$c_p = \gamma c_v ,$$
$$= 5.21 \times 10^3 \text{ J Kg}^{-1}\text{K}^{-1} .$$

The gas, by the way, is helium.

Experimental Problems

Question 1

In this experiment, the thermal expansion coefficient of aluminum, α, is to be measured. Two glass plates with polished flat surfaces are supported on a pedestal with the top plate raised slightly at one end where it is supported by an aluminum tube. The apparatus is viewed from above under sodium light ($\lambda = 589$ nm) and dark fringes are seen aligned across the plate. The aluminum tube may be heated using a forced air heat gun and its temperature may be taken using a thermocouple which has been imbedded in the base of the tube.

Question 2

In a simple transformer circuit, a power source with fixed voltage V_1 is connected across the primary windings and a variable resistive load R_L is connected across the secondary windings. The input power, P_1, and the power dissipated by the load, P_L, can both be measured. Plot P_L as a function of P_1 for varying values of R_L. Determine the maximum power dissipated by the load and the value of P_1 at which it occurs. Give your estimates of the errors attached to these values. These measurements illustrate the concepts of "impedance matching."

Question 3

A frequency generator can be regarded as a "black box" containing a voltage source and an effective resistance R_0. Using an oscilloscope as a voltmeter and a variable resistance box, determine R_0. Estimate the error in your results. Is the value of R_0 frequency dependent? If you wish to make a device where the output voltage is independent of the load resistance R, what must be the relation between R and R_0? If you wish to make a device where the output current is independent of the load resistance R, what must be the relation between R and R_0?

Question 4

A physical pendulum has the period

$$T = 2\pi \sqrt{\frac{I}{mgl}}$$

where I is the moment of inertia of the "bob" about the pivot, l is the distance between the centre of mass of the bob and the pivot, m is the mass of the bob and g is the acceleration due to gravity. Construct a pendulum from a quantity of thread and a tennis ball of radius r. Plot the square of the period of your pendulum versus l ranging from $l = r$ to $l \gg r$. What is the length of a simple pendulum which has the same period as your physical pendulum suspended such that $l = r$?

P. R. CHINA

INTERNATIONAL PHYSICS OLYMPIAD
AND
MIDDLE SCHOOL PHYSICS COMPETITION IN CHINA

Shen Ke-qi*   Zhao Kai-hua**
Peking University

1. The Chinese Olympiad Team well performed in IPhO

China joined the International Physics Olympiad (IPhO) for the first time in 1986 with an undermembered team (3 students). Since then a full Chinese team participated in the IPhO every year. The performance of Chinese team in IPhO in the last 4 years is listed in Tab.1. By the total score of the whole team China won the 3rd, 5th and 2nd place in 1987, 1988 and 1989, respectively.

Tab.1 The Performance of Chinese Olympiad Team in IPhO

	Students	Prize	Places
1986 IPhO XVII London, UK	1. LIN, Chen 2. WEI, Xing 3. ZHANG, Ming	II III honorable mention	9th 21st 33rd
1987 IPhO XVIII Jena, GDR	1. CHEN, Xun 2. LI, Jin-hui 3. TANG, Peng-fei 4. WU, Ai-hua 5. ZHANG, Yan-ping	II II III III III	4th tied for 11th tied for 29th tied for 37th tied for 37th
1988 IPhO XIX Bad Ischl, Austria	1. CHEN, Yan-song 2. XU, Jian-bo 3. CHEN, Feng 4. DING, Ai-dong 5.CHEN, Jian	I II II III honorable mention	4th 15th 23rd 54th 67th
1989 IPhO XX Warsaw, Poland	1. YAN, Jing 2. MAO, Yong 3. QIU, Dong-yu 4. GE, Ning 5. LIN, Xiao-fan	II II II II III	13th 16th 22nd 23rd 56th

*   Vice-president, CPS; Chairman, National Committee of MSPC.
**   Associate Sectretary-General, CPS; Member of ICPE.

It is not easy for Chinese team to achieve such an accomplishment among so many strong European teams. We would attribute our success, besides others, to the Middle School Physics Competition (MSPC) organized by the Chinese Physical Society (CPS). It was just this activity that had aroused wide enthusiasm for studying physics among middle school students and provided a broad and sound basis for selecting candidates to the Chinese Olympiad Team.

2. The Middle School Physics Competition in China

The history of science competitions in China has not been long. The MSPC has been held yearly only since 1984. It aims to enhance the interests and initiatives of students in studying physics, to have good effect on raising student's independent learning ability, to promote various extracurricular physics activities which enlivens the learning atmosphere of the school, and to find out the most talent middle school students in order to cultivate them more effectively.

A National Committee for MSPC is organized by CPS which is responsible for formulating the Competition Regulations. This Committee is composed of physics professors, middle school teachers and representatives from provincial Physical Societies. In each province, the Provincial Committee for MSPC is in charge of local competition affairs. This Competition is approved by the State Education Commission of P.R.China and supported by the local administrative education departments.

MSPC consists of two phases. The first phase is the preliminary competition which consists of written test and practical test. Every middle school student excellent in physics course can apply for participating in the written test. The number of participants each year is shown in Tab. 2,

Tab.2   Number of Total Participants in the Written Test of First Phase MSPC Each Year

1984	1985	1986	1987	1988
43 079	52 925	58 766	57 523	55 855

The written test papers are prepared by the National Committee and graded by the Provincial Committee. After the written test dozens of students with highest scores are selected to take practical test. The practical questions are prepared by the Provincial Committee, different for different provinces, and the test is conducted in the laboratories of the universities in the capital of each province. Both multiple-choice questions and free response questions are available in the written test. The provincial team participating in the final competition is organized through selection according to the overall scores of written and practical tests in the preliminary competition. Each provincial team is composed of three students in general, but for provinces with gold or silver award students in the last competition, one or two more students are allowed. The final competition, second phase of the national competition, is conducted by the National Committee. The number of participants is around one hundred. Every participant is required to take both written test and practical test. In addition to that, around a dozen of contestants with highest scores is required to take an oral exam which aims to select gold award winners from them. According to the Regulation, number of gold award, silver award and bronze award winners are around 5, 20, and 45 respectively.

MSPC proceeded successfully and reached its goals as mentioned above in the past years. Both the teachers and students in middle schools are interested in this extracurricular activity. The contestants with high scores were admitted to universities or colleges and exempted from College Entrance Examination. The candidates of Chinese Olympiad Team members are also selected from those students with highest scores in the final competition.

3. Training and Selection of the Chinese Olympiad Team

Around 15 students are selected to take an intensive training program. Such training is necessary since the syllabus of the middle school physics course in China is quite different from that of the IPhO. Part of the contents in the latter has not been taught in Chinese middle schools. The English level of the students can't meet the needs for communication with foreigners while going abroad. Therefore an intensive training of English is necessary as well. The period of training program is about 2 months. After the training, members of the Chinese Olympiad Team are selected finally.

The Chinese Association for Science and Technology(CAST), of which CPS is a member, is responsible for making contact with the organizer of IPhO as well as sending the Chinese Physics Olympiad Team abroad. CPS is responsible for selection of team members and intensive training with the help of a university. Besides, both CAST and State Education Commission support the Chinese Team financially.

APPENDIX IX

SOME SELECTED COMPETITION TASKS FROM MSPC IN CHINA

I. Theoretical Problems

Problem 1. Two gliders A, B collide on a linear horizontal air track. A stroboscopic photograph taken at the instant $t_0 = 0$, $t_1 = \Delta t$, $t_2 = 2\Delta t$, $t_3 = 3\Delta t$ is shown in the following figure. Determine when and where did the collision occur from the photo and data given.

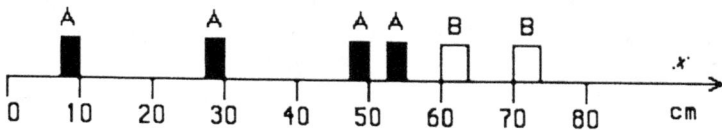

Solution The photo shows only 2 images of glider B in 4 strobo-shots, overlapping of images must have happened. Three possible cases could be assumed:

(1) Before collision the glider B moved to the left and left 2 images on the photo. After collision its motion turned to the right and gave 2 images coincident with the former images.

In this case, the coordinates of B are $x_{B0} = 70$, $x_{B1} = 60$, $x_{B2} = 60$, $x_{B3} = 70$, successively, so its speed is $10/\Delta t$ both before and after the collision. From this we have the collision time $t = 1.5 \Delta t$ and collision position $x_B = 55$, where there is an image of A shown on the photo. Since there was no strobo-shot at the moment of collision, glider A must have stand still either before or after the collision, both cases imply overlapping images for A. It is shown to be not the case by the photo.

(2) Before collision the glider B was at rest and left 3 coincident images on the photo. In this case collision occurred at position $x = 60$, i.e., $x_{B0} = x_{B1} = x_{B2} = 60$ and $x_{B3} = 70$. By the figure coordinates of A read $x_{A0} = 10$, $x_{A1} = 30$,

$x_{A2}=50$, $x_{A3}=55$, successively, from which the speed of A should be $20/\Delta t$ before collision and it collides with B at position $x=60$ and at instant $t=2.5\Delta t$. After collision, A moves to x_{A3} with speed $5/0.5\Delta t=10/\Delta t$, B moves to x_{B3} with speed $10/0.5\Delta t=20/\Delta t$. Then the consideration of momentum conservation gives $m_B=(3/2)m_A$, and simple calculation gives the values of total kinetic energy $E_k=200m_A/(\Delta t)^2$ and $E_k'=350m_A/(\Delta t)^2$ before and after collision, respectively, i.e., $E_k < E_k'$. This violates the conservation of energy, so this case must also be ruled out.

(3) The glider B was at rest after collision and left 3 coincident images on the photo. In this case the collision happened at $x=60$ and t between t_0 and t_1. By the figure $x_{B0}=70$, $x_{B1}=x_{B2}=x_{B3}=60$; $x_{A0}=55$, $x_{A1}=50$, $x_{A2}=30$, $x_{A3}=10$, successively. So the speed of A after collision is $20/\Delta t$, time of collision is $t=0.5\Delta t$, and then the speeds of A and B before collision are $10/\Delta t$ and $20/\Delta t$, respectively. From these data $m_B=(3/2)m_A$ by momentum conservation requirement. The corresponding values of kinetic energy are $E_k=350m_A/(\Delta t)^2$, $E_k'=200m_A/(\Delta t)^2$, $E_k > E_k'$, this is the only case possible to occur.

Problem 2. A circular ring of negligible thickness is put around the top of a cylindrical rod of length L and uniform cross-section as shown in the figure. The ring and the cylinder are of the same mass m. The ring is able to slide along the rod with a dynamical friction $kmg(k>1)$ which is equal to the maximum static friction, while the rod itself slide along a fixed vertical pole AB freely. Assume that there is no energy dissipation when the rod collides with the ground and the time of contact is extremely short. It is given that the rod falls freely from height H (counting from its lower end to the ground) and the ring slips away from its lower end after the nth bounce on the ground.

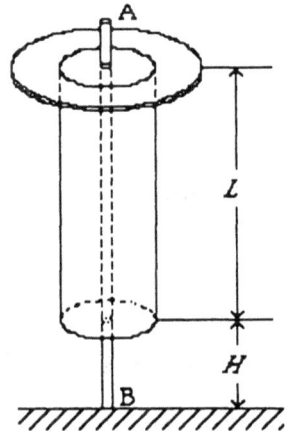

(1) Describe the motion of the ring and the rod in the process before the 2nd bounce. Find the height H' of the rod when the ring ceases to slide along it.

(2) Find the relation that should be satisfied by n, H, L and k.

Solution   (1) Quantities (velocity, acceleration, displacement) relevant to the first bounce are listed as following (taking downward as the positive direction):

quantities	rod	ring	ring relative to the rod
velocity just before collision	$\sqrt{2gH}$	$\sqrt{2gH}$	0
velocity immediately after collision	$v=-\sqrt{2gH}$	$\sqrt{2gH}$	$v'=2\sqrt{2gH}$
acceleration after collision	$a=(k+1)g$	$-(k-1)g$	$a'=-2kg$

If the ring keeps sliding along the rod, the time for the rod to reach its maximum height after first bounce is

$$t=-v/a=\sqrt{2gH}/(k+1)g=\frac{1}{k+1}\sqrt{\frac{2H}{g}}.$$

The time for the ring to cease sliding is

$$t'=-v'/a'=2\sqrt{2gH}/2kg=\frac{1}{k}\sqrt{\frac{2H}{g}}.$$

Obviously, $t' > t$, that means, the condition for calculating t· is satisfied. When the rod falls down from maximum height, the sliding of the ring will last a certain time interval. So long as the ring ceases to slide, the rod and ring will fall together with acceleration g. The total distance travelled by the ring along the rod in this bounce is

$$S_1=\frac{-(v')^2}{2a'}=\frac{2H}{k}.$$

At $t=t'$ the height of rod's lower end is

$$H'=-[vt'+a(t')^2]=\frac{H}{k}\frac{k-1}{k},$$

and the velocity of rod and ring at that moment is

$$v'=v+at'=\frac{1}{k}\sqrt{2gH}.$$

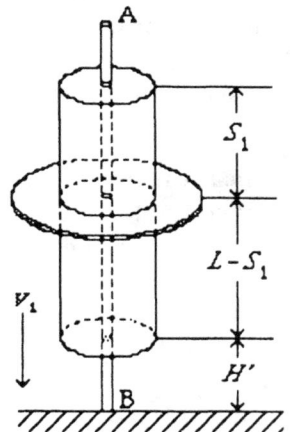

This is equivalent to the velocity of a body falling freely from rest at the height

$$H_2 = \frac{v_1^2}{g} + H' = \frac{H}{k} .$$

That is to say, for the 2nd bounce, it is equivalent to that the rod and ring fall from rest at height $H_2 = H/k$, so the formula of the sliding distance of the ring would be obtained simply by substituting H/k for H in the formula for the former one. This relationship holds for any two successive bounces so long as the ring does not slip away from the lower end of the rod, therefore the slicing distances after successive bounces will be

$$S_1 = \frac{2}{k} H ,$$

$$S_2 = \frac{2}{k} \frac{H}{k} = \frac{2H}{k^2} ,$$

$$S_3 = \frac{2}{k^2} \frac{H}{k} = \frac{2H}{k^3} ,$$

$$. . . . .$$

(2) During the process of the first $n-1$ bounces the total distance travelled by the ring along the rod is

$$S = S_1 + \cdots + S_{n-1} = \frac{2H}{k} \frac{1 - \dfrac{1}{k^{n-1}}}{1 - \dfrac{1}{k}} .$$

The condition for the ring to slide from the rod after the n th bounce is

$$S < L \leq S + \frac{2H}{k^2} ,$$

or

$$\frac{2H}{k} \frac{1 - \dfrac{1}{k^{n-1}}}{1 - \dfrac{1}{k}} < L \leq \frac{2H}{k} \frac{1 - \dfrac{1}{k^n}}{1 - \dfrac{1}{k}} .$$

Problem 3. Six resistors, identical in appearance, are connected as shown in the Figure. Their resistances are all equal to 2Ω except one which differs remarkably. Show that the exceptional resistor can be identified among the others by 3 measurements with an ohmmeter. Describe briefly your measuring method and reasoning.

Solution If there was no exceptional resistor, the effective resistance between any pair of the tetrahedral vertices would be equal to 1Ω. Measure the effective resistance between each pair of vertices of a triangle, say ABC, of the tetrahedral circuit. There are three measurements altogether. Then we can make analysis as follows. If the exceptional resistor lies on one of the sides, say AB, of this triangle, we must have

$$r_{BC} = r_{CA} \neq r_{AB}, \qquad r_{AB} \neq 1\,\Omega,$$

otherwise we should have

$$r_{CA} = r_{AB} \neq r_{BC}, \qquad r_{BC} = 1\,\Omega$$

if AD is the exceptional one.

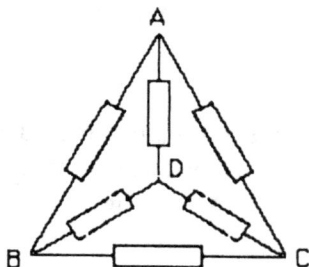

Problem 4. In the circuit shown in the Figure all resistances of the resistors are equal to 1Ω, the internal resistances of the ammeter and the cell are negligible, the terminal voltage applied by the cell is $10\,V$. Find

(1) the current in ammeter,

(2) the mean power output, if resistors r_1, r_2 and r_3 are now on, now off with probability $1/2$ and the changes of these 3 resistor are independent to each other.

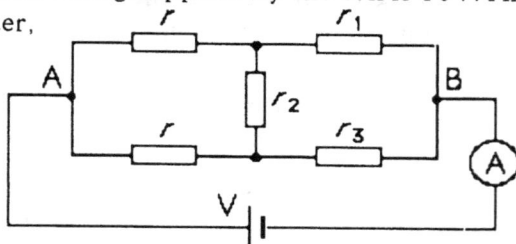

Solution (1) $10\,A$.

(2) There are 8 equally probable cases:

r_1	r_2	r_3	resistance between AB	power output
on	on	on	$1\,\Omega$	$100\,W$
on	on	off	$5/3\,\Omega$	$60\,W$
on	off	on	$1\,\Omega$	$100\,W$
on	off	off	$2\,\Omega$	$50\,W$
off	on	on	$5/3\,\Omega$	$60\,W$
off	on	off	∞	0
off	off	on	$2\,\Omega$	$50\,W$
off	off	off	∞	0

average value $= 52.5\,W$

II. Experimental Problems

Problem 1. Given a black box with 4 numbered terminals, inside which 3 elements are connected in a way unknown otherwise than that at most one element is connected across each pair of terminals, or they are simply shorted. The element may be one of the following: battery, resistor, capacitor, inductor, or diode.

(1) Determine what elements are used in the box and plot the circuit diagram. Give the inference of the uniqueness of your anwer.

(2) Determine the values of these elements. Describe briefly your measuring methods and cite the formulas needed for calculation.

Apparatus: Audio-frequency generator with variable frequency and output voltage, multimeter, connecting wires.

Solution The actual circuit in the black box is shown in the Figure. The student is required to obtain this circuit and the values of the resistance, inductance and capacitance of these elements through experimental measurements and analysis.

(1) Since there are at most three kinds of elements among the 5 listed kinds in the black box, one should identify which kinds of elements are existing at first. The following is an example of the possible procedures adopted.

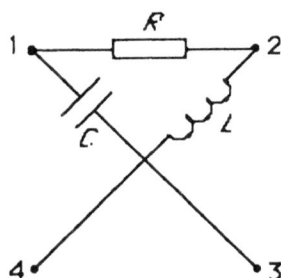

(i) The student should examine whether the battery exists or not at first by using multimeter as voltmeter.

(ii) Since diode resistance depends strongly upon the polarities of the loads, the existence of the diode can be easily justified by using ohmmeter.

Actually the possibility of existence of battery and diode is ruled out on this experiment.

(iii) Measure the resistance between every pair of terminals and observe if there is charge-discharge phenomena or not (which is characteristic for the existence of capacitors) simultaneously.

An example of the data obtained is shown in the following table.

pair of terminals	1,2	1,3	1,4	2,3	2,4	3,4
charge-discharge phenomena	no	yes	no	yes	no	yes
measured resistance value (Ω)	183	∞	200	∞	17	∞

One can reach the following conclusions through analysis of the above data:

a. Not short-circuited anywhere.

b. There is one and only one capacitor in the circuit, this capacitor is not connected across (1,2), (1,4) and (2,4).

c. Two other elements (resistor or inductor) are connected across (1,2) and (2,4) respectively.

d. There are only three possible cases as shown below,

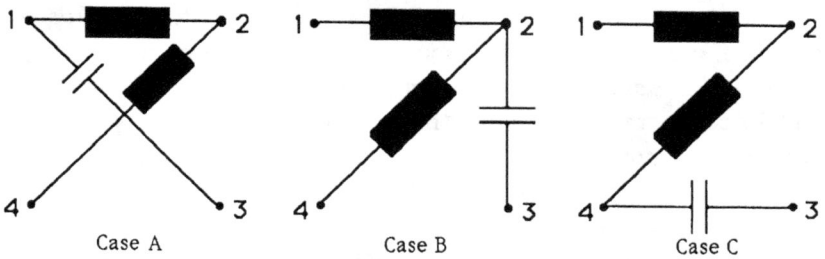

Case A   Case B   Case C

where the black bars denote either a resistor or an inductor.

(iv) Connect the signal generator to the terminals (3,4) and measure the voltage across (1,2), only the case A gives non-zero voltage, so case A can be discriminated from the others. Then measure the voltage across (2,4), the case C will give zero voltage while case B will give certain voltage value. Thus we can discriminate B from C. It is found that A is the actual case.

(v) Since the impedance of an inductor depends on frequency of the ac voltage applied and that of the resistor does not, so the resistor and inductor can be discriminated from each other by ac measurements

Now we get the circuit diagram as shown in the begining.

(2) Determine the characteristic values of these elements as follows. The resistance R can be measured by the ohmmeter.

Connect the signal generator output across the terminals 2, 3. Measure the voltage U_R and U_C by the ac voltmeter. The capacitance C is given by

$$C = \frac{U_R}{2\pi f\, U_C R}\ .$$

Similarly, connect the oscillator output across the terminals 1, 4, and measure the voltages U_L and U_R by the ac voltmeter, then the inductance is given by

$$L = \frac{U_L R}{2\pi f U_R}$$

if the resistance of the inductor can be neglected in comparison with $2\pi f L$.

Problem 2. When a small sphere of radius r moves at speed v in a fluid, the viscous drag exerting on the sphere is given by the Stokes' law:

$$f = 6\pi\eta r v, \tag{1}$$

where η is the coefficient of viscosity of the fluid, for which the standard unit in SI is pascal·second ($kg \cdot m^{-1} \cdot s^{-1}$).

Putting the sphere in a fluid, the viscous resistance increases with its falling speed. When the resistance plus the buoyancy balances the gravity:

$$\frac{3}{4}\pi r^3 \rho\, g = \frac{3}{4}\pi r^3 \rho'g\ + 6\pi\eta r v, \tag{2}$$

(ρ and ρ' are densities of the sphere and the fluid, respectively, g is the acceleration of gravity) a uniform terminal velocity v is attained. If ρ, ρ' and g are known, measuring r and v, one cancalculate η by the following formula derived from Eq.(2):

$$\eta = \frac{2\,(\rho - \rho')g\, r^2}{9\,v}\ . \tag{3}$$

Eq.(3) is based on the Stokes' law which applies, strictly speaking, only to the fluid of infinite extent. In the present experiment, however, castor oil in a long cylindrical container is used. To eliminate the error introduced by the wall effect, spheres with different radii are provided. Process appropriately your data obtained in the experiment with these spheres to get a more accurate value of η.

Apparatus Castor oil; a long cylindrical tube with 2 scratched marks at a distance l apart on its wall, the upper mark should be low enough to ensure that the sphere attains the terminal velocity before passing it;

stop-watch; micrometer calliper; 8 small ball bearings of 4 different radii (2 for each radius); screw-based tripod; plumb bob; jigs.

   Data available

   density of the balls r = 7.90 g/cm³,

   density of castor oil r ´= 0.950 g/cm³,

   acceleration of gravity g = 979.4 cm/s²,

   room temperature T = ----.

   Solution   (1) Fill the tube with castor oil and adjust itsverticality. Drop the sphere along its central axis.

   (2) Measure the diameters d of ball bearings by the micrometer calliper.

   (3) Measure the time of descending between 2 marks of 8 ball bearings. The experiment is repeated and the data for balls of the same size are averaged.

   (4) Calculate the values of η by substituting the data of $r = d/2$ and $v = 1/t$ for spheres of different sizes into Eq.(3). Plot a graph of η vs d, extrapolate it to the case of d =0, the intercept η_0 on the ordinate gives the value of the coefficient of viscosity free from wall-effect error.

COLOMBIA

THE PHYSICS OLYMPIAD IN COLOMBIA

Fernando Vega Salamanca
Luis Alejandro Ladino

The Universidad Antonio Nariño, in collaboration with the Ministry of National Education, has organized during the past few years the Mathematics, Physics and Computer Olympiads on the national level as one of its extension programs, has chosen students on the basis of these examinations to represent our country in international events such as the International Mathematical Olympiad, the Iberoamerican Mathematical Olympiad and the International Physics Olympiad, and has prepared these students academically for their participation in these international events.

Colombian students have 11 years of schooling divided in two parts: 5 years of primary school and 6 years of secondary school, the latter provides a four year basic cycle and a two year "superior" cycle. During these last two years the official curriculum includes 3 hours per week for physics courses. It is a very poor intensity, as can easily be observed, and this fault in official educational policy has not been corrected despite scholars attempts to do so. Also, in our country there is a marked difference in the quality of the education in the different social levels, between upper and lower classes, and between the cities and the provinces. Another aspect that influences the results of physics education in Colombia are the poorly equipped laboratories.

However, some schools on their own initiative increase the amount of physics classes weekly to four or the number of years that physics is taught to three or four.

In tenth grade the subjects included in the official cuuriculum are mechanics and thermodynamics; in eleventh grade harmonic motion, electricity, optics and some modern physics.

It is worth mentioning that in our country there are two different academic years, one which ends in December (A calendar) and another which ends in June (B calendar).

In 1985 the First Colombian Physics Olympiad (CPO) was organized. This first version contemplated only one level covering both 10th grade and 11th grade. The need for establishing two levels, because of the diversity of the programs, and for stimulating the study of each of the subjects in the programs, became immediately clear.

CURRENT ORGANIZATION OF THE OLYMPIAD

Participation. Students of tenth and eleventh grades, as well as students who have finished their studies during the first semester of the current year, may take part in the Colombian Olympiad. In general, schools register voluntarily for the olympiad, inscribing a certain number of their students. Individual registration of students whose schools did not register is also accepted.

Levels. We have two different levels, one for tenth grade and the other for eleventh grade, with the aim of providing incentives for study of each of the respective programs. In the examinations for the eleventh grade we include questions corresponding to the tenth grade program, encouraging students to appreciate the unity of physics as a science.

First Classifying Examination. Programmed especially for B calendar students, it is taken in mid-May. This first test consists of 20 multiple choice questions; the students write their answers on a card that can be read by computer. The grading of the exams takes place in Bogotá. Up to the present, the grading has been done as follows: every student starts with 20 points, for every correct answer he gets four points, for every incorrect answer -1 point and 0 points for every question not answered. This way scores between 0 and 100 can be obtained. The results are sent to each school.

Since every question has five answers, the probabilty of scoring 30 points, which is not enough to pass to the next exam, is a bit more than 4%. This practically assures objective results and, extending this to several hundred students, the possibility for a participant to continue if he is 'playing lottery' is eliminated by the second exam. On the other hand, the questions cover the entire program of each level which excludes the possibility of obtaining a good score for a student whose knowledge is based on his own interests.

Second Classifying Examination. This is given to the A calendar schools in September. The exam, as far as the number of questions, structure and grading are concerned, is similar to the First Classifying Examination. Even though the exams are programmed according to the different academic calendars, each school may participate in either or both of the exams. A student can classify for the following round of competition by the top score he obtains on either of the two exams.

Sellective Examination. This examination is scheduled one month after the second classifying examination. About 10% of the students who originally register will participate in this exam. Although the exam is devised in Bogotá by the organizing committee of the olympiad, it is taken in the different regions of the country. The exam consists of 10 questions which must be answered fully directly in the question booklet. The grading of this exam also takes place in Bogotá. The solutions and lists of results are sent to each school.

Physics Workshop. Five days before the Final Round of the olympiad, a physics workshop is organized with an intensity of six hours per day. The aim of this workshop is to give all participating students the chance to reach the same level, to help learning difficulties, and to teach how to approach certain problems. The sessions are designed to be interesting and agreeable to the students. There are some experimental demonstrations and exhibitions of scientific videos. Distinguished professors of various universities are invited to the workshop.

Final Round. The Final Round of competition takes place in the Graduate School of the Universidad Antonio Nariño in Bogotá at the end of the month of November. The number of students who take this test is about 10% of the number of students who take the sellective exam. It consists of a theoretical part and an experimental part. The theoretical exam consists of three or four questions which must be answered in 4½ hours, while its experimental counterpart contains one or two problems which must be solved in the same amount of time. The jury is composed of of five to seven people, some of them members of the organizing committee, and others university and high school professors. The grading is done in the following way: each problem is scored between 0 and 10, so the maximum score for the theoretical exam is 30 and for the experimental exam is 20. Each member of the jury grades separately giving a certain score for each of the questions. The grades are added and averaged.

The students who reach the final round are awarded a diploma by the Ministry of Education. The five best students of each level receive a gold medal. Diverse prizes and awards are given.

The members of the team that takes part in the IPhO are offered scholarships by various national universities.

About 4000-5000 student register for the classifying examination, including both levels. For the next round about 10% participates and in the final round the number of participants is about 50.

TRAINING OF THE TEAM FOR THE IPhO

The preparation for the internacional competition must be intensive, because the official curriculum is less extensive than the international program.

20 to 30 of the students which have participated in the final round are invited to the training session, which is divided in two parts. The preparation is done according to the international program. The first part takes place from February to May during which time the students must attend for four hours on Saturdays in Bogotá. Every month they are examined. The students who do not live in Bogotá may prepare differently,

but they must take the monthly examinations. The classes are basically theoretical.

According to the results of this first stage of training, 10 to 12 students are chosen. The second part of the training program runs for three weeks in the capital city of Bogotá, during which 6-8 hour daily sessions are programmed. Three times a week there are experimental practices. There are two examinations given per week. According to the results of these exams, the best students are chosen 10 days before the international olympiad. The students that are chosen must continue to train until the very last moment.

THEORETICAL PROBLEMS

1. Three identical cubes of side 1 m each are placed as shown in the figure 1:

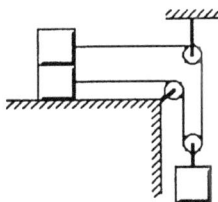

Fig.1

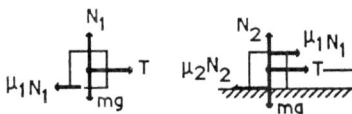

Fig 2

How long does the top cube take to separate from the bottom cube when the whole system starts moving from rest and the friction coefficient between the cubes is 0.2 and is 0.6 between the bottom cube and the table ?

A. 1/4 s   B. 2/3 s   C. 1/2 s

D. 2/5 s   E. 1/3 s.

Solution C The forces that are exerted on each body can be seen in the diagrams (see Fig. 2). Applying Newton's second Law to both the top and the bottom cubes, we have:

$$T-\mu_1 N_1 = ma_1, \quad N_1 = mg; \quad T-\mu_2 N_2 + \mu_1 N_1 = ma_2, \quad N_2 = 2mg$$

respectively.

with the values of N_1 y N_2 we get:

$$T - \mu_1 mg = ma_1, \quad T - 2\mu_2 mg + \mu_1 mg = ma_2$$

From the latter equations, by subtraction,

$$a_2 - a_1 = 2g(\mu_2 - \mu_1)$$

But this is the relative aceleration of the top body in relation to the bottom one, so that from the equation $e = at^2/2$, we obtain: $t = \sqrt{1/(\mu_2 - \mu_1)g}$.

2. A mattress is made of springs which are arranged by pairs as shown in Fig. 3. All of the springs have an identical elastic constant of 10 N/m. A mass of 100N weight is put on the mattress and as a result the level of the surface of the mattress goes down 10 cm. How many springs does the mattress have?. Assume that all of the springs are comprised the same lenght when the mass is put on them.

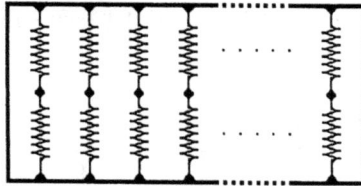

Fig. 3

A. 400   B. 100   C. 10

D. 70   E. 370

Solution A When we have two springs of elastic constants K_1 y K_2 placed in one case in serie and in parallel in the other case, the constants of the equivalent spring are given by $K_s = K_1K_2/(K_1+K_2)$ and $K_p = K_1 + K_2$ respectively. Then, the original system can be substituted by a system of n springs arranged in parallel with each spring having a constant equal to $K = k/2 = 5N/m$ as a result of having substituted the springs placed in serie just by only one. Then, if the springs satisfy Hook's Law we can write $F = nKx$, being n the number of double springs that now are in parallel. With the numerical data it turns out that $n = 200$, so that the mattress has 400 springs.

3. A block is able to slide down one of the inclined planes as shown in the figure 4. In one situation, it can move toward the left and in the other it can move toward the right. In both situations there is a friction coefficient

between the plane and the block. Once the body has stopped in each case as far as the reached distances AO and OB are concerned we can say that:

A. AO equal OB
B. OB greater than OA
C. OB lesser than OA
D. To answer, it is necessary to know the friction coefficient.
E. It is impossible to answer unless the angles of inclination of the planes are given.

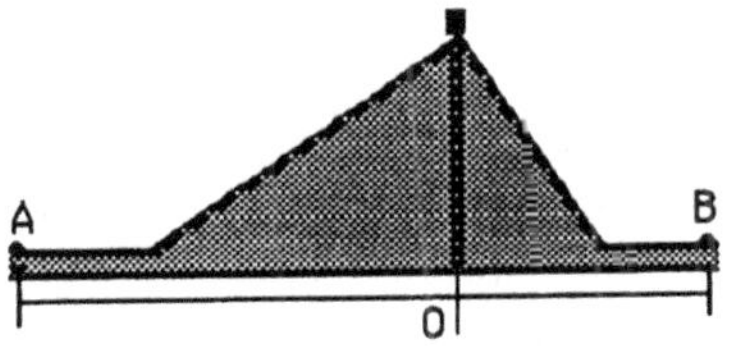

Fig. 4 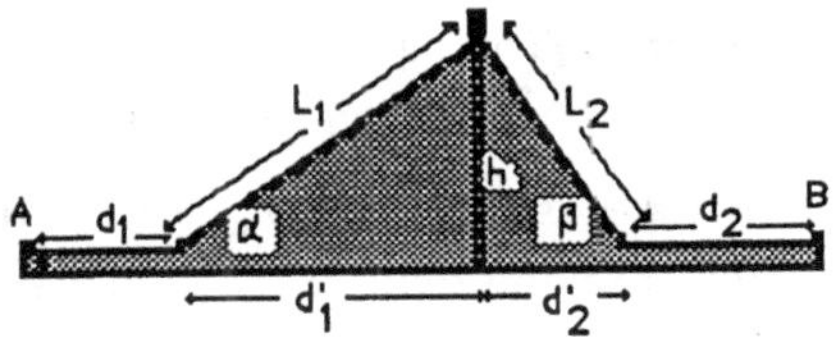   Fig. 5

Solution A. Taking into account the conservation of energy we can see easily that in whichever direction the body move, all the potential energy is spent as heat owing to the friction force and as work performed by the gravitational force. If the body moves toward the left, we have (Fig. 5): $mgh = \mu mg \cos\alpha\, L_1 + \mu mg d_1$ where $L_1 = h / \text{sen}\alpha$. Similarly, when the body moves toward the right: $mgh = \mu mg \cos\beta L_2 + \mu mg d_2$ where $L_2 = h / \text{sen}\beta$. Combining these two expressions we can write: $d_1' + d_1 = d_2' + d_2$.

4. Let the following system (Fig. 6) be, which consists of a rod AB. The mass M can move along the rod freely. The left piston has a radii three times greater than the right one. Determine the ratio ℓ_1/ℓ_2 so that the rod AB stands horizontally.

A. 1/9   B. 1/3   C. 2/5
D. 1/4   E. 2/3.

Solution A. Taking the position of the mass as a rotation axis and writing the moments according to this, the condition of equilibrium is $pS_1 l_1 = pS_2 l_2$.

5. Two tubes of diameters 4 cm and 3 cm are connected to each other as shown in Fig. 7, inside the tubes there is some water. How much would the level of the water increase if we put a cylinder made of wood with a diameter of 1 cm and height of 25 cm in the left tube?. The density of the wood is $2\ g/cm^3$.

> **A.** 2 cm   **B.** 1 cm   **C.** 1.5 cm
>
> **D.** 2.4 cm   **E.** 1.8 cm

Solution B. The equilibrium condition of the block is:

$$\frac{\pi}{4}D_1^2\,\rho_0 g\Delta h=\frac{\pi}{4}D_m^2\rho_m\,gh-\frac{\pi}{4}D_2^2\,\rho_0 g\Delta h\ .$$

Where D_1, D_2, ρ_0, D_m and ρ_m are the diameters of the tubes, the density of the water, and the diameter and density of the wooden block respectively. From this equation we can obtain the required height:

$$\Delta h=\frac{D_m^2\,\rho_m\,h}{(D_1^2+D_2^2)\rho_0}\ .$$

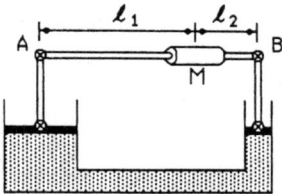 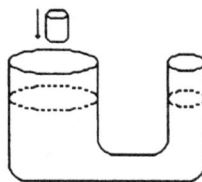 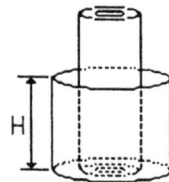

Fig. 6   Fig. 7   Fig. 8

6. A hollow wooden cylinder of height h is placed in a cylindrical container (Fig. 8). The internal and external radii of the cylinder and the container are R, 2R, 3R respectively. Determine the minimun height of the container so that when we pour water into it the cylinder can float inside freely.

> **A.** h   **B.** $h\rho_{water}/(\rho_{water}+\rho_{wood})$
>
> **C.** h^2/R **D.** $h\,\rho_{wood}/\rho_{water}$
>
> **E.** $h[(R+R\,(\rho_{water}/\rho_{wood})]/R$.

Solution D. When the body is floating it is in its equilibrium position so that, it must satisfy: $\Sigma\ \vec{F_i}=0$. The weight of the body is given by P =

$3gR^2hg\pi\rho_{wood}$ which is equal to the Arquimedes' force that is given by ρ_{water} $3gR^2Hg\pi$, being H the minimun height of the container. Equating these two expressions it turns out $H = (\rho_{wood}/\rho_{water})h$.

7. A charge Q_2 can move freely between two charges Q_1 and Q_3 which are fixed and placed on a straight line, the charge Q_2 stays in the position shown in figure 9a. Then Q_2 and Q_3 are fixed and we let the charge Q_1 move freely to the position shown in figure 9b. Find the value of each of the charges if it is known that the smallest value of the charges is 1C. Suppose that all the charges are positive.

A. 1 C, 4 C, 8 C B. 1 C, 16 C, 4 C C. 1 C, 2 C, 8 C
D. 1 C, 16 C, 8 C E. 1 C, 8 C, 16 C

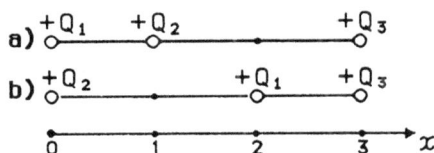

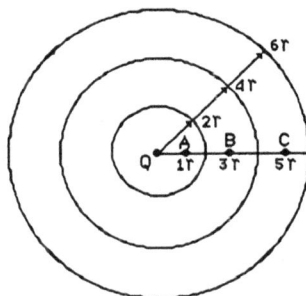

Fig. 9 Fig. 10

Solution B Let us denote by $\vec{F}_{ij}$ the force that the charge i exerts upon the charge j. Then for the first situation it must satisfy that $|\vec{F}_{12}| = |\vec{F}_{32}|$, which it is equivalent to $Q_1Q_2 = kQ_2Q_3/4$, according to Coulomb's law. Similarly it turns out for the second situation $kQ_1Q_2/4 = kQ_1Q_3$. Solving these pair of equations,we get: $Q_3 = 4Q_1$ and $Q_2 = 4Q_3$. Now, because the smallest charge is 1C, it must be that this correspond to Q_1, in this way $Q_1 = 1C$, $Q_2 = 16C$ and $Q_3 = 4C$.

8. Let a configuration of concentric metalic surfaces be as shown in Fig. 10. A charge Q is put at the center of the spheres. What must be the value of the charge that has to to be put on each surface so that the intensity of the electric field be the same at the points A, B and C?

A. 4Q, 8Q, 12Q   B. Q, 4Q, 9Q   C. 8Q, 16Q, 24Q

D. 9Q, 27Q, 81Q   E. Q, 9Q, 81Q.

Solution C. Inside a conducting spherical surface the value of the electric field generated by charges placed on the center itself is null. Note that the charge on the third surface can have any value. According to the given conditions at points r, 3r and 5r we have,

$$k\frac{Q}{r^2} = k\frac{Q}{9r^2} + k\frac{Q'}{9r^2} = k\frac{Q}{25r^2} + k\frac{Q'}{25r^2} + k\frac{Q''}{25r^2}$$

where Q', Q'' are the charges of the spherical surface of radii 2r and 4r respectively. From these relations we get the required charges.

9. We have two test tubes without any gauge in their side walls. It is known that the first has a volume of 8 ml and the second one has a capacity of 8ml. Propose a method to gauge these test tubes in ml.

Solution Let us denote the test tubes of 8ml and 5 ml capacities by A and B respectively. First of all we fill B, then this liquid is poured into A, in this way we can make the "5ml" mark on it. Once again, we fill B and with this liquid we proceed to fill A, so that there are 2ml left in B and we can make "2ml" mark on it. Now we take the liquid out of the test tube A and pour the 2 ml from B to it and proceed to make the "2ml" mark on it. Then, we fill B and we pour it into A, so to the 2ml that were in A we add 5ml, in this way we proceed to make the "7ml" mark and so on. The procedure can be represented by the diagram in figure 11, following the direction of the arrows: In the x-axis we place the capacity of A and in the y- axis the capacity of B.

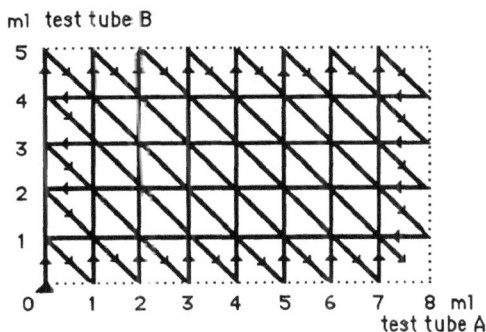

Fig. 11

10. A cottage is built by five square sheets of side L m (Fig.12). It is known that a drop of water falls over the roof of the house takes twice the time to go from point A to point B than to go from point B to point C. Find the time that the drop takes to go from point A to point C. Suppose that the drop started moving from point A with cero initial velocity.

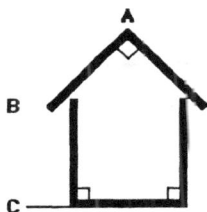

Fig. 12

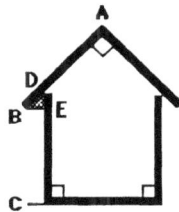

Fig. 13

Solution From the shadowed triangle BDE (Fig.13) we see that

$$X = BE = ED = \frac{\sqrt{2}\,L - L}{2} = \frac{(\sqrt{2} - 1)L}{2}. \tag{1}$$

with,

$$h = L - X = \frac{(3 - \sqrt{2})L}{2}. \tag{2}$$

The drop crosses the distance $\overline{AB}$ in

$$2t = \sqrt{2L/a} \tag{3}$$

from which we find

$$a = 2L/t^2 \tag{4}$$

taking into account the given conditions, the distance h is crossed in t or,

$$h = vt + gt^2/2, \tag{5}$$

where v is the vertical component of the total velocity in B. Taking into account that

$$v = \cos 45° \; v_{total} = \sqrt{aL},$$

and putting the value of (4) into it and then the value of v obtained like this is put into (5). The value that we get for h is put into (2) so, t is equal to

$$\sqrt{(3 - 2\sqrt{2})\frac{L}{g}}.$$

The total time to cross the distance AC is 3t.

11. Two bodies of identical mass are tied by a cord which passes through a pulley, as shown in fig. 14. The friction coefficient between the bodies and planes is the same. What is the minimun value of the angle θ that makes the system start moving?

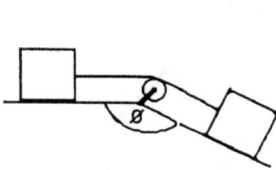
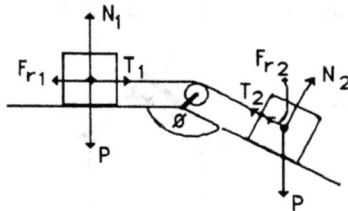

Fig. 14   Fig. 15

Solution The forces that act upon the bodies are shown in the figure 15. The equilibrium condition for body 2 is:

mkg + mkg cos (180° - θ) = mg sen (180° - θ)

Symplifying

$$k\,[1 + \cos(180° - \theta)] = \text{sen}\,(180° - \theta),$$

$$k\,(1 - \cos\theta) = \text{sen}\,\theta$$

by squaring,

$$k^2\,[1 - \cos\theta]^2 = \text{sen}^2\,\theta = 1 - \cos^2\theta,$$

hence,

$$(1 - k^2)\cos^2\theta - 2k^2\cos\theta + k^2 - 1 = 0.$$

solving this second degree equation, we obtain that

the first root is: $\cos \theta_1 = 1$, or $\theta_1 = 0°$, which has no physical sense. The second root

$$\cos \theta_2 = \frac{k^2 - 1}{k^2 + 1}$$

so

$$\theta_2 = \arccos \frac{k^2 - 1}{k^2 + 1}.$$

12. A vehicle carries out a maneuver as shown in the graph of velocity against time (Fig. 16). If the work done by the motor is cero. What is the friction coefficient?

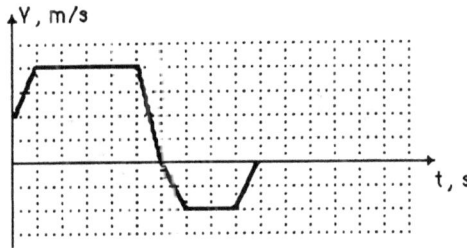

Fig. 16

Solution Let us denote by A_t^i, A_m^i, A_f^i the total work, and the work performed by the motor and friction force respectively. Here the superindice indicates the corresponding sector. It is clear that $A_t^i = A_m^i + A_f^i$ and their values for the different sectors according to the diagram are

$$A_t^1 = m(4^2 - 2^2)/2,$$
$$A_t^2 = 0,$$
$$A_t^3 = - m4^2/2,$$
$$A_t^4 = m(-2)^2/2,$$
$$A_t^i = 0, A_t^i = - m(-2)^2/2.$$

The total work

$$A_t = \sum A_t^i = 0 + (-kmgS),$$

where $S = \sum S^i$ is the path travelled by the vehicle. Using the average velocity for each sector v^i, we can write

$$S^i = v^i t = (v_i + v_f)t/2$$

Hence

$$S^1 = 3 \text{ m,}$$
$$S^2 = 16 \text{ m,}$$
$$S^3 = 2 \text{ m,}$$
$$S^4 = 1 \text{ m,}$$
$$S^5 = 2 \text{ m,}$$
$$S^6 = 1 \text{ m,}$$

or S = 25 m. Therefore, k = 2/(gS) = 2/250 = 0,008.

13. The system shown in the figure 17 consist of two pullies, a spring with an elastic constant k and a hook that has a mass of two tons. How much must the spring be compressed so that the cord becomes an equilateral triangle?

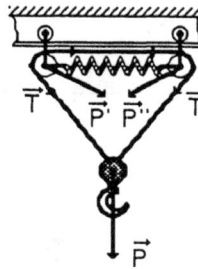

Fig. 17   Fig. 18

Solution The forces that are exerted on the system are shown in the diagram (see Fig.18). We have assumed that the hook has a mass much greater than that of the cord. Due to the symetry it is easy to see that P' = P'' = P. It must be taken into account that the forces that act upon the cord are equalled by the force that the cord exerts on the hook. Because the cord becomes an equilateral triangle, the force applied to one end of the spring is

F = P'cos 30°

Considering Hook's law, we have

$$\Delta x = \frac{F}{k} = \frac{P'\cos 30°}{k} = \frac{mg\cos 30°}{k} ,$$

with the numerical values

$$\Delta x = 0.0069 \text{ mm.}$$

14. There are four billiards balls on a straight line which is parallel to one side of the table, the balls are 1 m apart (Fig. 19). The ball A is hit in such way that it acquires a speed of 50 cm/s. Find the minimum time so that the balls return to the initial configuration. Explain your answer. Assume the collisions are completely elastic.

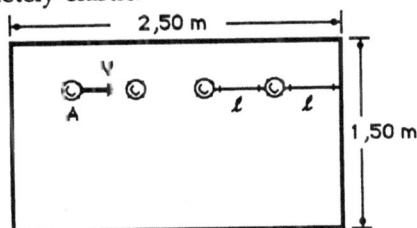

Fig. 19

Solution The collision between the balls is completely elastic, so that the conservation of momentum and energy can be written as

$$m\vec{v} + m\vec{0} = m\vec{v_1} + m\vec{v_2}$$

$$\frac{mv^2}{2} - \frac{m0^2}{2} = \frac{mv_1^2}{2} + \frac{mv_2^2}{2}$$

Symplifying,

$$v = v_1 + v_2,$$
$$v^2 = v_1^2 + v_2^2.$$

The only acceptable solution is $v_1 = 0$ and $v_2 = v$, which it is equivalent to say that the first ball stays at rest after the collision and the second ball acquires the velocity of the first one. Thus, as the four balls are identicals and the collision of one ball against the edge of the table is elastic (at this point the ball bounces backward with the same magnitude of velocity), we may say that every time that a ball collides against another ball, we are going to have the initial situation. Therefore, the time taken by a ball to travel the distance ℓ is $t = \ell/v$. Considering all collisions that happened, we can readily see that the total required time is 8 seconds.

15. A bus A moves forward with velocity v and collides against a bus B that was at rest (Fig. 20). On the top of bus B there is suitcase as shown in a). After the collision both buses continue moving together until they stop. All this happened within a second. As a result of the collision the suitcase moves from bus B to bus A. The suitcase is 1 m long and its mass is 140 times smaller than the mass of any of the buses, the friction coefficient between the suitcase and the surface of the bus is 69 times greater than the friction coefficient between the buses and the road. Find the velocity v.

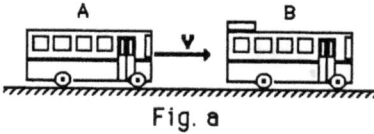

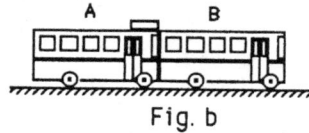

Fig. a   Fig. b

Fig. 20

Solution The diagram shows the forces that act upon the bodies along the horizontal direction (see Fig. 21) (Both buses are considered as one body).

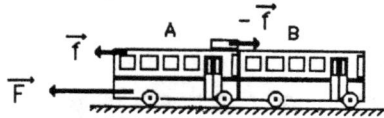

Fig. 21

According to the Newton's second Law

$$\vec{F} + \vec{f} = 2M\vec{A}$$
$$-\vec{f} = m\vec{a}$$

Where $\vec{F}, \vec{f}, M, m$ are the friction forces between the buses and the road and between the suitcase and the buses, the mass of each bus and the mass of the suitcase respectively.

From the conservation of momentum

$$M\vec{v} = 2M\vec{V} + m\vec{w}$$

where v,V and W are the initial velocity of the A bus, the velocity of the buses after the colission and the velocity of the suitcase. All the previous velocities are measured relative to the earth.

Writting the same equations in a scalar way, we have,

$$Mv = 2MV + mw,$$
$$-F - f = 2M(-A),$$
$$f = ma.$$

Replacing the friction forces by their numerical values

$$Mv = 2MV + mw,$$
$$[(2M + m) k_1 + mk_2] g = 2MA,$$

$$mk_2g = ma.$$

Acording to the given conditions: $M = 140\ m$ and $k_2 = 69\ k_1$ we have

$$140\ v = 280\ V + w,$$
$$350\ k_1g = 280\ A, \tag{1}$$
$$69\ k_1g = a.$$

Because the bodies stop at the end, we have finnally the following cinematic relations

$$0 = \vec{V} + \vec{A}t, \quad \text{or} \quad 0 = V - At \tag{2}$$
$$0 = \vec{w} + \vec{a}t, \quad \text{or} \quad 0 = w - at.$$

From the equation (1) we get v, and dividing the second of this equation by the third one, we have

$$v = 2V + \frac{w}{140} \quad \text{and} \quad \frac{A}{a} = \frac{5}{4(69)}$$

From the equations (2) we have,

$$V = \frac{wA}{a} = \frac{5w}{4(69)}. \tag{3}$$

16. There is an ideal gas inside a test tube with a small diameter d (Fig. 22). The gas is separated from the air by a mercury drop of mass m=2g. What is the ratio L_2/L_1? Being L_1 the length taken by the gas when the test tube is turned upwards and L_2 the length when the test tube is upside down.

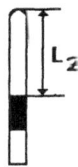
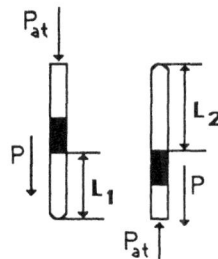
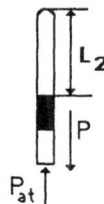

Fig. 22   Fig. 23

Solution Regarding the procces as isothermical, the state equation for the ideal gas is

$$PV = \text{const.}$$

Thus,we can relate the parameters of the gas for the two situations in the next way

$$P_1V_1 = P_2V_2.$$

Now, considering the pressures that act upon each volume of gas, we have (Fig. 23)

Obtaining:

$$\frac{V_2}{V_1} = \frac{P_1}{P_2} = \frac{P_{at} + \dfrac{P}{S}}{P_{at} - \dfrac{P}{S}}$$

where S is the internal cross section area of the test tube and P is the weight of the mercury drop. Using P = mg, we get

$$\frac{V_2}{V_1} = \frac{SL_2}{SL_1} = \frac{P_{at} + \dfrac{mg}{S}}{P_{at} - \dfrac{mg}{S}}$$

eliminating S

$$\frac{L_2}{L_1} = \frac{P_{at} + \dfrac{4mg}{\pi d^2}}{P_{at} - \dfrac{4mg}{\pi d^2}}$$

17. We have three identical copper cubes A, B and C respectively. The cube A is at 200°C and the other two are at 0°C. Would it be possible to achieve that the cube A acquiered a temperature lower than the other two cubes? What would the method be? (the interchange of heat with the enviroment is neglected)

Solution Let us denote the state of the bodies by $(t_1 \degree C, t_2 \degree C, t_3 \degree C)$ where each cypher denotes the temperature of each body.

Thus, the inicial state is (200 °C, 0 °C, 0 °C). Now, putting together bodies A and B during enough time to reach equililibrium, we have

$$mc\,(200 - t\degree) + mc\,(0 - t\degree) = 0,$$
$$2t\degree = 200 \degree C,$$

where c, m are the heat capacity and the mass of each. Hence, $t\degree = 100 \degree C$.
Thus, the second state is (100 °C, 100 °C, 0 °C).
Then, putting together the bodies A y C until they reach equilibrium, we have,

$$mc\,(100 - t\degree) + mc\,(0 - t\degree) = 0,$$

hence, $t° = 50 °C$.

Thus, the third state is (50 °C, 100 °C, 50 °C).

Similarly we proceed to do the same, but this time with B and C, obtaining

$$mc (100° - t°) + mc (50 - t°) = 0,$$

hence, $t° = 75 °C$.

And we have the required state (50 °C, 75 °C, 75 °C).

18. A ring shaped tube contains three ideal gases of identical mases and molar masses M_1, M_2, M_3 respectively. The gases are separated by a stopper that can move freely without any friction (Fig. 24). What are the values of the angles α_1, α_2, α_3, formed by the stoppers

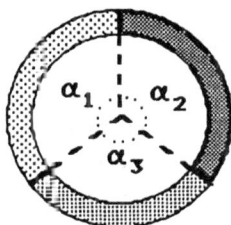

Fig. 24

Solution The state equation of each one of the gases is given by:

$$\frac{P_1 V_1}{T_1} = \frac{m}{M_1} R, \qquad \frac{P_2 V_2}{T_2} = \frac{m}{M_2} R \quad y \quad \frac{P_3 V_3}{T_3} = \frac{m}{M_3} R .$$

Dividing the first and third equation by the second and taking into account that the temperature and the pressure are the same for all them, we have

$$\frac{V_1}{V_2} = \frac{M_2}{M_1} \quad y \quad \frac{V_3}{V_2} = \frac{M_2}{M_3} .$$

The angle is related to the corresponding volume by $V = k\alpha$, so that $V_1 = k\alpha_1$, $V_2 = k\alpha_2$ and $V_3 = k\alpha_3$. Obtaining

$$\frac{\alpha_1}{\alpha_2} = \frac{M_2}{M_1} \quad y \quad \frac{\alpha_3}{\alpha_2} = \frac{M_2}{M_3}$$

224

but

$$\frac{\alpha_1+\alpha_2}{\alpha_2}=\frac{M_2+M_1}{M_1}$$

and adding α_3

$$\frac{\alpha_1+\alpha_2+\alpha_3}{\alpha_2}=\frac{M_2+M_1}{M_1}+\frac{M_2}{M_3}.$$

It is clear that $\alpha_1+\alpha_2+\alpha_3=360°$, then

$$\alpha_2=360°\frac{M_1 M_3}{M_1 M_2+M_1 M_3+M_2 M_3}.$$

similarly

$$\alpha_1=360°\frac{M_2 M_3}{M_1 M_2+M_1 M_3+M_2 M_3},$$

$$\alpha_3=360°\frac{M_1 M_2}{M_1 M_2+M_1 M_3+M_2 M_3}.$$

19. A cilindrical container is sealed by a piston that is held by a spring as shown in Fig. 25. At the begining the container is emptied, then, a certain amount of gas is put inside at certain temperature and this gas is heated until it reaches a temperature equal to twice the initial temperature. Prove that the first increment of the volume cannot be equal to the second increment of it. Neglect the effect of the external atmotsphere, the mass of the piston and also the friction with the walls.

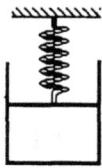

Fig. 25

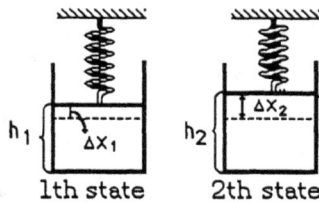

1th state   2th state

Fig. 26

Solution The diagrams (Fig. 26) shown the position of the pistons when the gas is inside. The state equation for both states is

$$\frac{P_1 V_1}{T_1}=\frac{P_2 V_2}{T_2}$$

but,

$$T_2 = 2T_1,$$

Putting in the previous equation the values of the pressures and volume, we have,

$$2 \frac{F_1}{S} h_1 S = \frac{F_2}{S} h_2 S$$

where S, h_1 and h_2 are the area of the bottom of the cylinder, the heights reached by the piston in the first and second case.
Hence,

$$2F_1 h_1 = F_2 h_2.$$

Putting the values of the forces given by Hook's Law

$$2k\Delta x_1 h_1 = k\Delta x_2 h_2,$$
$$2\Delta x_1 h_1 = \Delta x_2 h_2.$$

Its clear from the figure that

$$2\Delta x_1 h_1 = (\Delta x_1 + \Delta h)(h_1 + \Delta h),$$

where $\Delta h = h_2 - h_1$. Multiplying both sides of this expression by S^2, we have

$$2\Delta V_1 V_1 = (\Delta V_1 + \Delta V)(V_1 + \Delta V).$$

Now, according to the supposition, if the relation $\Delta V = \Delta V_1$ is fulfilled we would have

$$2\Delta V V_1 = 2\Delta V (V_1 + \Delta V),$$

hence, $\Delta V = 0$. which it is impossible.

20. A cylindrical tube with its piston is placed in a container with water as shown in Fig. 27, a column of air h m high and with a atmospherical pressure of p_0 is between the piston and the water surface. Then the piston rises b m from the water level inside the tube. Calculate the height of the water column inside the tube knowing that the height of the liquid column in the water barometer at atmospherical pressure is c m.

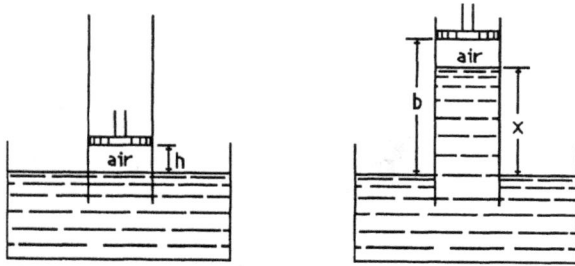

Fig. 27

Solution The state equations for the left and right gases are,
$$P_0 Sh = nRT \qquad PS(b-x) = nRT$$
being S the cross-section area of the piston. Note that we have put the same value of the temperature because we have assumed that the gas doesn't change its temperature during the process.

Equating these two expressions
$$gCh = P(b-x)$$

Besides, we have that $P_0 = P + gh$, where $P_0 = gC$. Putting this expression in the latter one we have one quadratic equation

$$X^2 - (c+b)X + c(b-h) = 0$$

where,
$$X = \left(\frac{b+c}{2}\right) \pm \sqrt{\frac{(b-c)^2}{4} + hC}$$

21. Is it possible to combine two resistors to make a heater in four different ways in such a way that we can obtain four different scales that give off quantities of heat equal to nQ_0 with n = 1,2,3, and 4? Explain your answer.

Solution The possibles combinations that can be done with two resistors R_1 and R_2 are four

    a) In parallel

    b) Only the resistor R_1

    c) Only the resistor R_2, and

    d) In serie.

As the combination a) gives off four times more heat than the combination d) we have,

$$\frac{u^2 t}{\frac{R_1 R_2}{R_1 + R_2}} = 4 \frac{u^2 t}{R_1 + R_2} ,$$

hence

$$(R_1 + R_2)^2 = 4 R_1 R_2,$$
$$(R_1 - R_2)^2 = 0$$

Thus, $R_1 = R_2$, in which case it is not possible to have four combinations.

EXPERIMENTAL PROBLEMS

22. Describe the possible ways of finding the density of a piece of deformed gray metal.

Discuss yours results and compare them each other.

Elements.

-A graduaded ruler

-A ligth wooden bar

-Two plactic glasses

-A light cord

-A piece of metal of unkown density

-One copper cylinder of density 8300 kg/m^3

-Water.

Before you start to work experimentally, you have to solve the problem analitically and write it down on the papers for each one of the methods you suggest. Once you have obtained them, try to do a diagram of the experimental set that you are going to use and explain the reason for such a set.

Further questions.

a) Explain the possibles sources of errors in your results.

b) How would your results be afected If you used a liquid different from water?.

c) Would your results be afected If you used a hollow cylinder instead of a solid one?.

d) How would your answer changed if we added some sugar to the water?
e) How would your answer be changed if we worked with inches instead of centimeters ?.

Solution. Let us consider the following two experimental sets (Fig. 28)

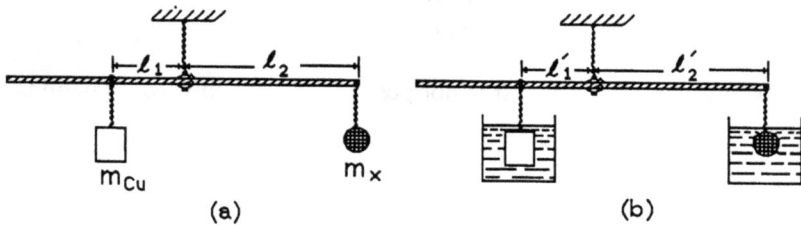

(a)   (b)

Fig. 28

At equilibrium the sum of the torques is equal to cero, so for the left system
$$m_{Cu}gl_1 = m_x gl_2 \tag{1}$$
being m_{Cu} and m_x the masses of the cylindrical piece of copper and the deformed unknown piece respectively. Now, from the right set and making use of the Arquimedes' principle it turns out

$$(m_{Cu}g - V_{Cu}\rho_{H_2O}g)\ell'_1 = (m_x g - V_x\rho_{H_2O}g)\ell'_2 \tag{2}$$

where ρ, V_{Cu} and V_x are the density of the water, the volumes of copper and the unknown mass pieces. In the previous equations the volume of the mass put at the right is unknown, and so is its mass. From $\rho = m/V$, we get

$$V_{Cu}\rho_{Cu}\, g\ell'_1 = V_x\rho_x\, g\, \ell'_2 \tag{3}$$
$$V_{Cu}(\rho_{Cu} - \rho_{H2O})\ell'_1 = V_x(\rho_x - \rho_{H2O})\, \ell'_2 \tag{4}$$

respectively. Finding ρ_x from (3) and putting it in (4), we have

$$V_x = V_{Cu}\frac{\rho_{Cu}}{\rho_{H_2O}}\cdot\left[\frac{\ell_1}{\ell_2} - \frac{\ell'_1}{\ell'_2}\right] + V_{Cu}\,\frac{\ell'_1}{\ell'_2} \tag{5}$$

The next step is to determine the mass m_x, to do it, we use the relation (1) because we can determine the mass of the copper piece from its density and volume, the volume is measured directly.
So

$$\rho_x = m_x / V_x$$

ALTENATIVE SOLUTION

Using the following experimental set (see Fig. 29)

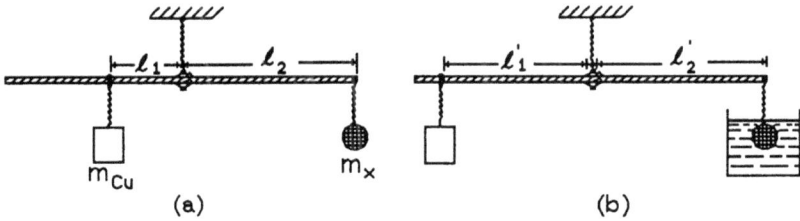

Fig. 29

we see that this set is similar to first, we can use the relations (1) and (3). Applying the same principles, we get

$$m_{Cu}\, g\, \ell_1' = (m_x g - V_x\, g \rho_{H_2O})\, \ell_2' \qquad (6)$$

$$V_{Cu}\rho_{Cu}\ell_1' = V_x(\rho_x - \rho_{H_2O})\ell_2' \qquad (7)$$

Finding ρ_x from (3) and putting it in (7)

$$V_x = \frac{\rho_{Cu}}{\rho_{H_2O}}\left[\frac{\ell_1}{\ell_2} - \frac{\ell_1'}{\ell_2'}\right] \qquad (8)$$

The rest of procedure is identical to the first

CYPRUS

THE PARTICIPATION OF CYPRUS

IN THE INTERNATIONAL

PHYSICS OLYMPIADS

by C. M. POURGOURIDES, Ph.D.

Headmaster, Athienou Secondary School.

C. KYRIAKIDES, M.Sc

Dassoupolis Lyceum.

1. INTRODUCTION.

Cyprus became interested in the International Physics
Olympiads in 1985 but the first participation was achieved
in 1988 at the 19th I. Ph. O. In view of our limited
experience in this field and in anticipation of the
establishment of the University of Cyprus (commencing in
1992) parts of the procedure for the training of the team
described in this article is liable to change.

2. THE SELECTION OF THE CYPRUS TEAM.

The Cyprus team is selected after a National
Competition which was established in 1987. The competition
is named "Pancyprian Physics Olympiad" and is carried out
by the Cyprus Science Association in cooperation with the
Ministry of Education of Cyprus.

2.1. Description of the Competition

The competition is open to all citizens of the Republic of Cyprus under the age of twenty. It is made known to interested persons through the press, the Cyprus Broadcasting Corporation, and by means of circulars sent to all upper secondary schools in the island at least three weeks before its commencement, which is set in the early spring. The competition is held at the four major cities of Cyprus, i.e Nicosia, Limassol, Larnaca and Paphos. The problems are set by a joint committee of physicists appointed by the Cyprus Science Association and the Ministry of Education. The language of the papers is Greek and English but provisions exist for translating the papers into other Cypriot languages such as Turkish or Armenian provided the candidate notifies the joint comittee one week in advance. At this stage the examination is theoretical and the grading is done at the examining centres. Each centre is required to send about seven of its best papers to the joint committee and a total of about 30 successful candidates is selected.

At the second stage the 30 candidates are required to take a second examination at a single examining centre at Nicosia. The examination is again theoretical but an experimental examination is to be introduced at this stage in the future. These candidates are graded and are all awarded certificates of participation according to their grede. The top ten are selected as members of the Cyprus team.

3. TRAINING OF THE TEAM BEFORE THE IPhO

The training of the team consists in:

(a) Covering additional syllabus which is required for the Olympiads and is not contained in the Cyprus Secondary School Physics curriculum such as Relativity, Thermodynamics and Quantum Physics.

(b) Solving theoretical problems from past Physics Olympiads and additional harder examples.

(c) Performing experiments to gain skill in the use of instruments and error analysis. The training is done at the Nicosia Paedagogical Institute. It is on a voluntary basis. Both teachers and students devote a lot of their free time, usually in the afternoons or during holidays for preparation. The teachers responsible for the preparation are those that will accompany the team to the country of the IPhO. Teachers and students do not gain any formal advantages for their extra work the only motivation being their love for Physics and the International Olympiads spirit.

The top five students distinguish themselves during this training and up to now it was not found necessary to conduct another competition for the final selection of the five candidates.

The expenses for travelling etc. to the country where the competition takes place are paid by the Ministry of Education. A smaller amount to cover extra needs is paid by the Cyprus Science Association.

It must however be noted that since our participation in the Olympiads, there has been a growing interest in Physics

in our country and a recognition of the importance of
international communication and exchange of experiences and
knowledge in the scientific field at this level.

3. THEORETICAL PROBLEMS GIVEN AT THE NATIONAL COMPETITION.

3.1. Problem 1:

Part (a) A long horizontal string AB of linear density
$\mu = 0.10 Kgm^{-1}$ is under tension $F = 40N$. The end A
executes S.H.M in a vertical direction with
amplitude $y_o = 5 \times 10^{-2} m$ and frequency 2Hz.
Find:

(i)   The velocity and wavelength of the waves
travelling along the string.

(ii)   The maximum acceleration of every element
of the string in the vertical direction.

(iii) The maximum vertical force experienced by a
length of the string $l = 0.5 \times 10^{-2}.m$

Part (b)   At two points S_1 and S_2 on a liquid surface two
coherent wave sources are set in motion with the
same phase. The speed of the waves in the liquid
$v = 0.5 ms^{-1}$ the frequency of vibration $f = 5Hz$ and
the amplitude $y_o = 0.04m$.
At a point P of the liquid surface which is at a
distance $x_1 = 0.30m$ from S_1 and $x_2 = 0.34m$ from S_2
a piece of cork floats.

(i)   Find the displacement of the cork at $t = 3$ s.

(ii)   Find the time t_o that elapses from the
moment the wave sources were set in motion
until the moment that the cork passes through
the equilibrion position for the first time.

3.2. Solution to Problem 1:

Part (a)

(i)   The speed of the wave $v = \sqrt{\dfrac{F}{\mu}}$

$$v = \sqrt{\frac{40N}{0.10 Kgm^{-1}}} \implies \underline{v = 20\ ms^{-1}}$$

the wavelength $\lambda = \dfrac{v}{f} = \dfrac{20\ ms^{-1}}{2\ s^{-1}} \implies \underline{\lambda = 10m}$

(ii) The maximum acceleration $a = -\omega^2 y_o \implies a = 4\pi^2 f^2 y_o$

$$\underline{a_{max} = 7.9\ ms^{-1}}$$

(iii) Neglecting the effects of gravity the maximum vertical force required is

$$F_{max} = m.a_{max} \implies F_{max} = \mu \ell a_{max}$$

$$F_{max} = 0.10 \times 0.5 \times 10^{-2} \times 7.9 \implies \underline{F_{max} = 3.95 \times 10^{-3} N}$$

If we take into account the effect of gravity,

$$F_{max} = ma_{max} + mg \implies F_{max} = \mu \ell (a_{max} + g)$$

taking $g = 10 ms^{-2}$

$$F_{max} = 0.10 \times 0.5 \times 10^{-2} (7.9 + 10)$$

$$\underline{F_{max} = 8.95 \times 10^{-3} N}$$

Part (b)

(i)   Let S_1 and S_2 be the two point sources and P the point under consideration. Since the speed of the waves is $v = 0.5ms^{-1}$ the time required for the wave to travel

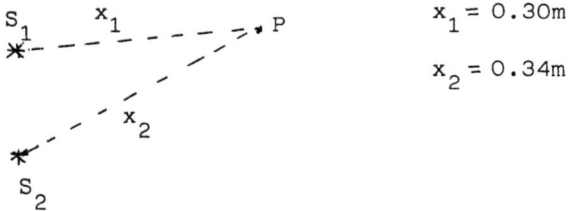

$x_1 = 0.30m$

$x_2 = 0.34m$

the distance S_1P is t_1 where

$$t_1 = \frac{x_1}{v} \Rightarrow t_1 = \frac{0.30}{0.50} \Rightarrow \underline{t_1 = 0.60 \text{ s}}$$

Also the time t_2 for the distance S_2P is

$$t_2 = \frac{x_2}{v} \Rightarrow t_2 = \frac{0.34}{0.5} \Rightarrow \underline{t_2 = 0.68 \text{ s}}$$

The time quoted $t = 3$ s is obviously greater than either t_1 or t_2. Concequently the displacement from the equilibrium position will be $y = y_1 + y_2$ where $y_1 = y_0 \sin 2\pi f (t - \frac{x_1}{v})$

$$y_2 = y_0 \sin 2\pi f (t - \frac{x_2}{v})$$

and

$$y = 2y_0 \cos \frac{\pi f (x_1 - x_2)}{v} \sin 2\pi f \cdot (t - \frac{x_1 + x_2}{2v})$$

Substituting we obtain

$$y = 2 \times 0.04 \cos \frac{5\pi (0.34 - 0.30)}{0.5}$$

$$\times \sin 2\pi \times 5 (3 - \frac{0.34 + 0.30}{2 \times 0.5})$$

$y = 0.08 \cos (5\pi \times 0.08) \sin 10\pi(3 - 0.64)$

$y = 0.08 \cos 0.4\pi \sin 23.6\pi \Rightarrow y = -0.235$ m

(ii) The piece of cork passes through the equilibrium position at times when its displacement is zero. These are given by the zeros of the phase term. Consequently

$$\sin 2\pi f \left(t - \frac{x_1 + x_2}{2v} \right) = 0$$

or $\sin (10\pi t - 6.4\pi) = 0$

the solutions to this equation are those values of t such that

$$10\pi t - 6.4\pi = K\pi \qquad K = 0, 1, 2 \ldots$$

which gives

$$t = \frac{K + 6.4}{10}$$

However we cannot start from K=0 since we must take into acount the time required for the wave to reach P. The required time t_x must obviously satisfy

$$t_x > t_1 \quad \underline{and} \quad t_x > t_2$$

·Hence $t_x > 0.68s \implies \frac{K + 6.4}{10} > 0.68$

or $K > 0.4$ the values of K start from $K = 1, 2 \ldots$ etc. The cork passes through the equilibrium position for $K = 1$ for the first time.

$$t_x = \frac{1 + 6.4}{10} \implies \underline{t_x = 0.74 \ s}$$

3.3. Problem 2.

Part (a)

A ring of radius r and cross sectional area A made of metal of density d and resistivity ρ is allowed to fall in a radial magnetic field of magnetic flux density B as shown in the fig. I and II.

If at a certain instant the velocity of the ring is v find the relation giving (i) The induced current in the ring
(ii) The acceleration of the ring.

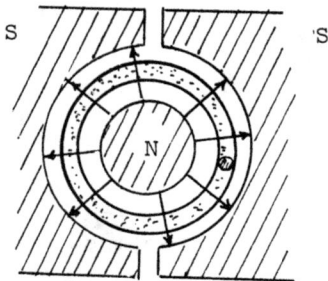

Fig. I

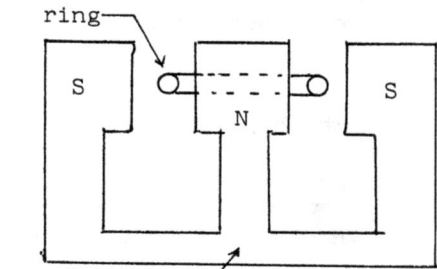

Fig. II

Part (b)

A metallic ring with its axis vertical is placed at a height y over an electromagnet supplied with a current $I = I_0 \cos \omega t$. If M is the mutual inductance (for this position of the ring) between the electromagnet and the ring and R the resistance of the ring.

(i) find the relation giving the induced current in the ring.

(ii) explain which will be the the direction of the resultant electromagnetic force acting on the ring.

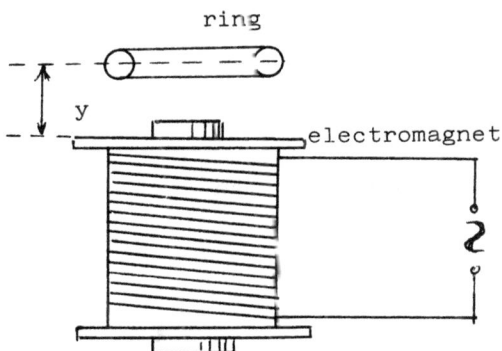

3.4 Solution to Problem 2.

Part (a)

(i) Consider an elementary segment $\delta\ell$ of the ring. At the moment when the speed is v, the induced E.M.F. between the ends of $\delta\ell$ is $\delta E = B.\delta\ell.v.$ For the whole ring

$$E = \Sigma\,\delta E = \Sigma B v\,\delta\ell = Bv\Sigma\delta\ell = Bv.2\pi r$$

r denotes the mean radius. The induced current I is

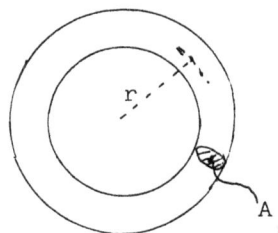

$$I = \frac{\text{Induced E.M.F.}}{\text{Total Resistance}}$$

$$I = \frac{2\pi r B v}{R}$$

Where R the total resistance of the ring. Neglecting inductance effects

$$R = \frac{\rho\ell}{A} \qquad \text{where} \quad \ell = 2\pi r \quad \text{and A is the cross}$$

sectional area

Hence $I = \dfrac{2\pi rBv}{\rho.2\pi r/A}$ $\Rightarrow$ $I = \dfrac{BAv}{\rho}$

(ii) The electromagnetic force on the ring $F = BI\ell$

$$F = B.\frac{BAv}{\rho}.2\pi r \quad \text{and since it tends to}$$

oppose the downward motion it acts upwards. From Newton's Second Law of motion

$$\Sigma \vec{F} = m\vec{a} \quad \text{we obtain}$$

$$mg - F = ma \quad \text{or} \quad a = g - \frac{F}{m}$$

$$a = g - \frac{B^2 v 2\pi r}{m\rho}$$

but the density $d = \dfrac{mass}{volume}$

$$d = \frac{m}{2\pi r \ A} \qquad \text{Substituting we obtain}$$

the acceleration

$$a = g - \frac{B^2 Av.2\pi r}{\rho.d.2\pi r \ A}$$

$$\text{or} \quad a = g - \frac{B^2 v}{\rho d}$$

Part (b)

(i) Let Φ denote the total magnetic flux through the ring. From the definition of the mutual inductance, $M = \dfrac{\Phi}{I}$ where $I = I_o \cos\omega t$, we obtain

$$\Phi = MI_o \cos\omega t.$$

The induced E.M.F is

$$E = -\frac{d\Phi}{dt} = -\frac{d}{dt}(MI_o \cos\omega t)$$

$$E = \omega MI_o \sin\omega t$$

and the induced current is now

$$I \quad \frac{MI_0\omega\sin\omega t}{R}$$

(ii) In an electromagnet
the value of the magnetic
flux density B decreases
as we move away from it.
With the ring at a
particular position,
B can be resolved into
two components: $B_y = B\sin a$ and
$B_x = B\cos a$

The direction of flow of I is as shown.
The component B_y is responsible
for electromagnetic forces such
as F_x which act along the
radius of the ring and by
symmetry they cancel out.
The component B_x generates
electromagnetic forces such
as F_y perpedicular to the
plane of the ring, vertically
upwards which add up.
The resultant elecromagnetic
force is vertically upwards.
This must be so because by
Lenz's law the force must be
such that it opposes the downward
motion of the ring.

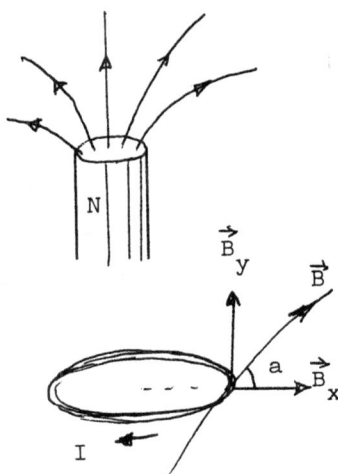

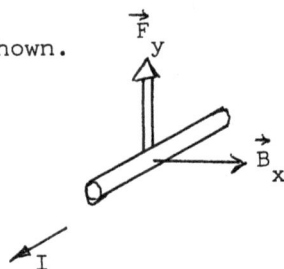

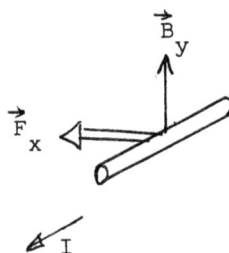

3.5. Problem 3

Part (a)

A light elastic string of natural length $AB = \ell_o = 1m$ is assumed to obey Hooke's law $F = Kx$. The string hangs freely from a fixed point A. When a small bob of mass $m = 0.2$ Kg is attatched at the free end B, the string is extended by OB (static extension) and comes to rest at the equilibrium position O. The bob is further depressed to the point C where $OC = 0.10m$ and when released, it is observed to excecute simple harmonic motion in the vertical direction with a period $T = 2$ s

Find:

   (i)   The modulus of elasticity of the string.

  (ii)   The velocity of the particle at the point D where
           $OD = 0.05m$.

 (iii)   The time required for the particle to move from
           C to D.

  (iv)   The maximum kinetic energy of the body.

Part (b)

   (i) The particle of mass m at the end of the string is
       raised back to point A and it is allowed to fall
       freely. Find the time taken for the particle to
       return to the point A for the first time.

(ii) Give a graph of the velocity of the particle against time for the motion of the particle described in (i)

3.6. Solution to Problem 3

Part (a)

(i) The period of oscillation $T = 2\pi\sqrt{\dfrac{m}{K}}$

$$K = \frac{4\pi^2 m}{T^2} \implies K = \frac{4\pi^2 \times 0.2}{2^2} \implies K = 2\,N/m$$

(ii) The velocity in S.H.M is $v = \omega\sqrt{x_0^2 - x^2}$ where $x_0 = OC\ 0.10$ m and $x = OD = 0.05$ m

$$\omega = \frac{2\pi}{2}\,s^{-1} \quad \text{or} \quad \omega = \pi\,s^{-1}$$

$$v = \pi\sqrt{0.10^2 - 0.05^2} \qquad \underline{v = 0.27\,ms^{-1}}$$

(iii) The time t_1 to move from O to D is given by

$x = x_0 \sin\omega t$ or

$0.05 = 0.10\sin\pi t$

$\sin\pi t = \dfrac{1}{2}$ $\qquad \pi t = \dfrac{\pi}{6}$

or $t = \dfrac{1}{6}$ s

The time required to move from O to C is $\dfrac{1}{4}$ of the period i.e $\dfrac{1}{4} \times 2 = \dfrac{1}{2}$ s. The required time is

$t_x =$ time OC − time OD

$$t_x = \frac{1}{2} - \frac{1}{6} = \frac{1}{3}\,s \qquad t_x = \frac{1}{3}\,s$$

(iv) The maximum kinetic energy is

$$E_{max} = \frac{1}{2}mv_{max}^2 = \frac{1}{2}kx_0^2$$

$$E_{max} = \frac{1}{2} \times 2 \times 0.10^2$$

$$E_{max} = 0.01 \text{ Joule}$$

Part (b)

(i) When the particle is raised back to the point A and released let L be the lowest point ite reaches when it comes to rest momentarily and let K be the position of final (static) equilibrium

Let $BK = x'$

$\quad KL = x'_o$

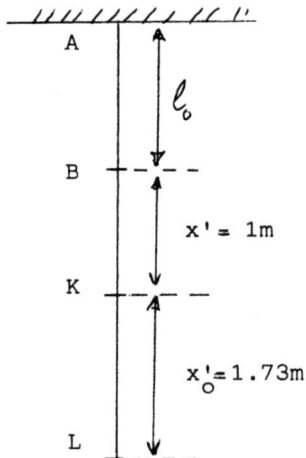

To calculate x' use $mg = Kx'$ $\Rightarrow$ $x' = \frac{mg}{K}$ $\Rightarrow$ $x' = \frac{0.2 \times 10}{2}$

$\underline{x' = 1 \text{ m}}$

To calculate x'_o use the principle of conservation of energy.

Potential Energy at A = Elastic potential energy at L

$$mg(\ell_o + x' + x_o) = \frac{1}{2} K (x' + x'_o)^2$$

$$0.2 \times 10(1 + 1 + x_o) = \frac{1}{2} \times 2 (1 + x'_o)^2$$

which gives $\quad \underline{x'_o = \sqrt{3} \text{ m}}$

The required time that elapses until the body reaches back to A again is denoted by

$$t = 2 (t_{AB} + t_{BK} + t_{KL})$$

where:

t_{AB} is the time required to fall freely through AB.

$$t_{AB} = \sqrt{\frac{2h}{g}} \quad \Rightarrow \quad t_{AB} = \sqrt{\frac{2\ell_o}{g}} \quad \Rightarrow \quad t_{AB} = \sqrt{\frac{2 \times 1}{10}}$$

$t_{AB} \quad 0.447 \text{ s}$

t_{BK} is the time required to move from B to K as it excecutes S.H.M.

$x' = x'_o \sin \omega t$

$1 = 1.73 \sin \omega t$

$\sin \pi t = \dfrac{1}{1.73} \quad \Rightarrow \quad t = 0.196 \text{ s}$

and $t_{BK} = 0.196$ s.

$t_{KL} = \dfrac{T}{4} \quad \Rightarrow \quad t_{KL} = 0.5$ s.

$t = 2 \ (0.447 + 0.196 + 0.5)$

$t = 2.286$ s

(ii) The required graph is shown below

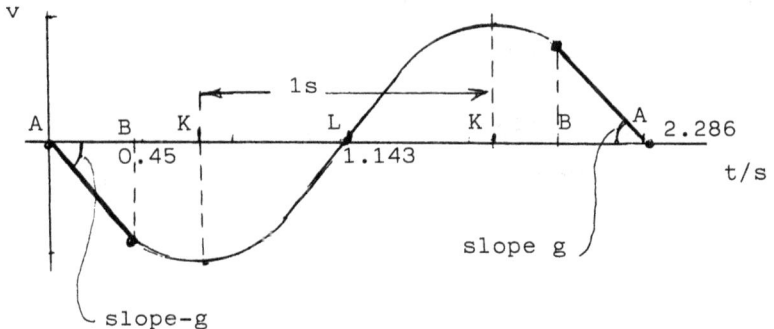

3.7   Problem 4:

Part (a)

(i)   A parallel plate capacitor is constructed from two
      square plates each of side 15 cm situated 5 cm apart
      in air ($\varepsilon_o = 8.85 \times 10^{-12}$ F/m)
      Calculate its capacitance.
      The capacitor is fixed with
      its plates vertical on
      insulating supports.

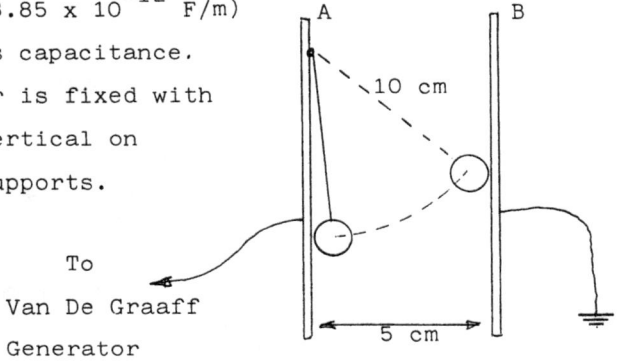

                To
          Van De Graaff
          Generator

(ii) A small, spherical pith ball coated with conducting
      paint hangs from a piece of silk thread of length
        10 cm the upper point of which is fixed on to the
      plate A.   The ball initially touches A.
      The pith ball has a mass $m = 0.1$ g and a radius
      $r = 0.3$ cm.
      Calculate the capacitance of the pith ball.

Part (b)

      Plate B is earthed and plate A is momentarily
      connected to a Van de Graaf generator charged to a
      potential of 60 000 volts. The plates are again
      isolated and it is observed that the pith ball moves
      from plate A to B and back to A several times until it
      finally comes to equilibrium with the thread hanging
      at an angle θ to the vertical.

(i)   Explain the behaviour of the pith ball and determine

its position of final equilibrium.

(ii) Calculate the final potential difference between the plates.

(iii) Find how many times (K) the ball goes back and forth before coming to rest.

(iv) Sketch a graph of the potential difference between the plates as a function of the mumber of journeys of the pith ball between the plates.

$$V_{AB} = f(K)$$

3.8 Solution to Problem 4:

Part (a)

(i) The capacitance of the parallel plate capacitor is

$$C = \frac{A\varepsilon_0}{d} \implies C = \frac{(15 \times 10^{-2})^{-2} \times 8.85 \times 10^{-12}}{5 \times 10^{-2}}$$

$$C = 3.98 \text{ pF}$$

(ii) The capacitance of the spherical pith ball is

$$C' = 4\pi\varepsilon_0 r \implies C = 4\pi \times 8.85 \times 10^{-12} \times 0.3 \times 10^{-2}$$

$$C = 0.334 \text{ pF}$$

Part (b)

(i) The pith ball is charged and is repelled by plate A. It touches B and is discharged. Gravity brings it back to contact with A and so forth. The pith ball transfers

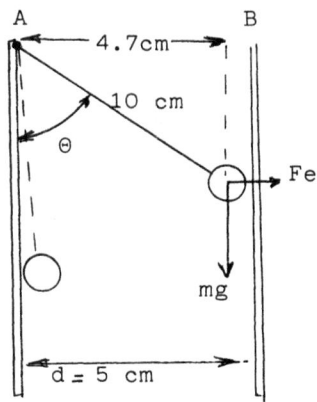

charge from A to B. Finally it comes to rest when it just fails to touch B. The thread makes an angle Θ with plate A where

$$\sin \Theta = \frac{4.7}{10} \qquad\qquad \Theta = 28^{\circ}$$

Θ is just less than 28°

(ii) The final potential difference is such that the electrostatic force on the pith ball and the force of gravity hold it at the angular position Θ

$$\tan \Theta = \frac{F}{mg} \quad\Longrightarrow\quad \tan \Theta = \frac{qE}{mg} = \frac{C'V_f^2}{dmg}$$

where $V_f =$ final potential difference

$$V_f = \sqrt{\frac{mgd.\tan \Theta}{C'}}$$

$$V_f = \sqrt{\frac{0.1\times10^{-3}\times9.81\times5\times10^{-2}\times \tan 28^{\circ}}{3.34 \times 10^{-13}}}$$

$$\underline{V_f = 8836 \text{ volts}}$$

(iii) During the first contact the amount of charge on the sphere is q_1 where $\quad q_1 = C'V_o$

V_o is the initial potential of plate A

$V_o = 60\ 000$ volts.

As the sphere moves away the potential of the plate A drops to V_1 because its charge is reduced from its initial value Q_o to Q_1. We have

$$Q_o = q_1 + Q_1$$

$$CV_o = C'V_o + CV_1$$

$$V_1 = \frac{CV_o - C'V_o}{C} \quad \text{or} \quad V_1 = V_o \left(\frac{C-C'}{C}\right)$$

After the second contact.

$$V_2 = V_1 \left(\frac{C-C'}{C}\right) \implies V_2 = V_o \left(\frac{C-C'}{C}\right)^2$$

After the ball has travelled K times

$$V_f = V_o \left(\frac{C-C'}{C}\right)^K$$

Assuming the maximum value of V_f to be 8836 volts then

$$8836 = 60\ 000\ \left(\frac{3.98 \times 10^{-12} - 3.34 \times 10^{-13}}{3.98 \times 10^{-12}}\right)^K$$

$$0.1473 = \left(\frac{39.8 - 3.34}{39.8}\right)^K$$

$$K = \frac{\log 0.1473}{\log 0.9161} \implies K = 21.8$$

Let us assume $K = 21$ and evaluate the final potential difference and the angle Θ

$$K = 21 \qquad \log \frac{V_f}{V_o} = K \log 0.9161$$

$$V_{21} = 9526.9 \text{ volts}$$

and $\tan \Theta_{21} = \frac{C'V_{21}^2}{dmg} \implies \tan \Theta = \frac{3.34 \times 10^{-13} \times (9526.9)^2}{5 \times 10^{-2} \times 0.1 \times 10^{-3} \times 9.81}$

$$\tan\Theta = 0.618 \implies \underline{\Theta = 31.7^{\circ}}$$

for $K = 22$ we obtain

$$\log \frac{V_f}{V_o} = 22 \times \log 0.9161$$

$$V_{22} = 8727.6 \text{ volts}$$

and $\tan\theta$ 0.519 $\Longrightarrow$ $\theta = 27.4^\circ$

This is less than 28°. The final potential difference is thus 8727.6 volts, the final angle 27.4° and the number of journeys 22.

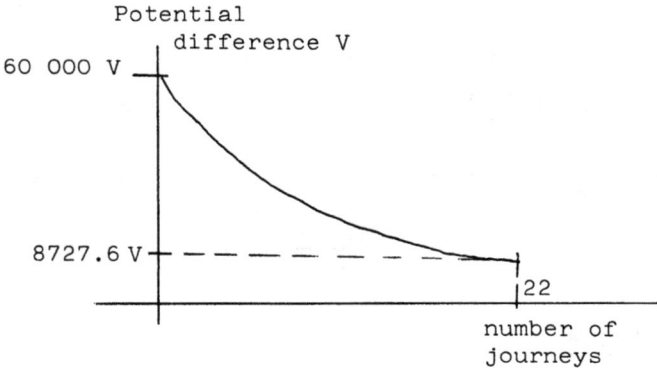

3.9. Problem 5

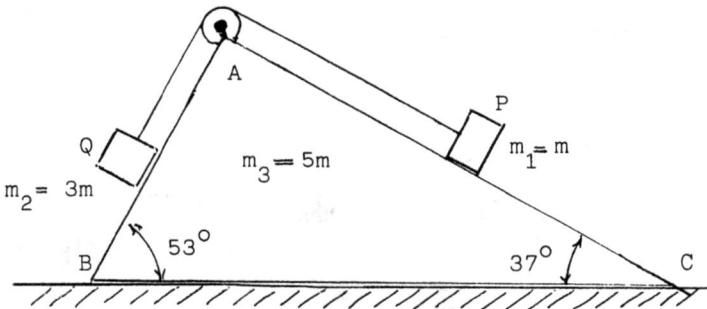

Two bodies P and Q of masses $m_1 = m$ and $m_2 = 3m$ respectively are joined by a light inextensible string. The string passes over a light, frictionless pulley fixed at the top corner A of a wedge of mass $m_3 = 5m$. The wedge has a triangular cross section ABC with angles $ABC = 53°$ and $ACB = 37°$ and rests with its face BC on a smooth horizontal table. BC can slide on the table. All three bodies are held stationary and then released at the same instant.

(i) Draw the forces acting on each body.

(ii) Assuming all surfaces frictionless, calculate the acceleration of each one of the three bodies with respect to the table surface assummed motionless.

(iii) Assuming friction between wedge and table surface calculate the value of the coefficient of friction so that the wedge just remains motionless on the table surface.

You may assume $g = 10 \ ms^{-2}$

$\sin 37° = 0.6 = \cos 53°$

3.10. Solution to Problem 5.

(i)

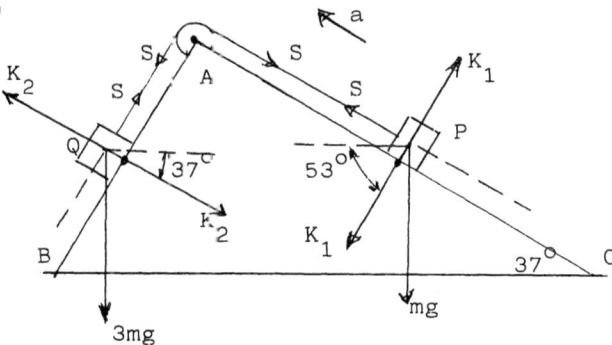

(ii)   Let a be the acceleration of P and Q with <u>respect</u> <u>to the block</u> along the direction AC assummed positive in the direction shown

$$a = \frac{\Sigma \vec{F}}{\Sigma m} \implies a = \frac{3mg\sin 53^{\circ} - mg\sin 37^{\circ}}{4\ m}$$

$$\implies a = 4.5 ms^{-2}$$

The tension S is found from $3mg\sin 53^{\circ} - S = 3ma$

$$S = 3m(g\sin 53^{\circ} - a)$$

$$\underline{S = 10.5\ m\ N}$$

The reactions are $K_2 = 3mg\cos 53^{\circ}$

$$K_1 = mg\cos 37^{\circ}$$

At the top A the two tensions S combine to give the action of the pulley on the block

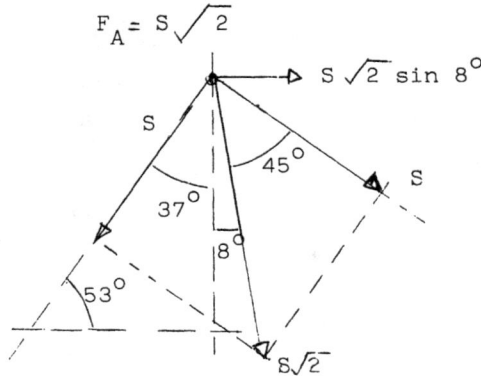

at an angle of 8° to the vertical. The component $S\sqrt{2}\ \sin 8^{\circ}$ contributes to the motion of the block. Let $a_3 =$ acceleration of block.

$$K_2 \sin 53^{\circ} + S\sqrt{2}\ \sin 8^{\circ} - K_1 \sin 37^{\circ} = 5m.a_3$$

$$a_3 = \frac{3mg\cos 53^{\circ}\ \sin 53^{\circ} + 10 \times 5m\sqrt{2}\ \sin 8^{\circ} - mg\cos 37 \sin 37^{\circ}}{5m}$$

$$a_3 = \frac{2 \times 10 \times 0.8 \times 0.6 + 1.46\sqrt{2}}{5} = 2.333 \text{ ms}^{-2}$$

The accelerations relative to the table top are denoted by a_1 and a_2 and are calculated below.

For body Q

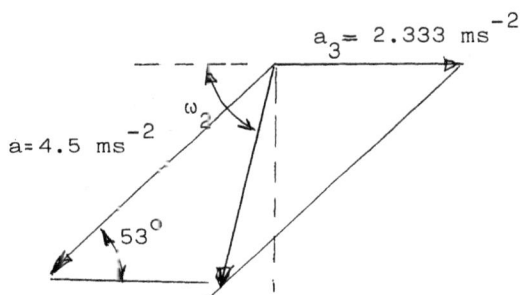

$a = 4.5 \text{ ms}^{-2}$

$a_3 = 2.333 \text{ ms}^{-2}$

ω_2

53°

$a_2 = 3.62 \text{ ms}^{-1}$; $\qquad \omega_2 = 84.2^\circ$

For body P

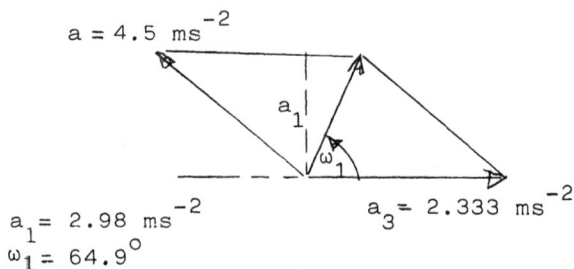

$a = 4.5 \text{ ms}^{-2}$

a_1

ω_1

$a_3 = 2.333 \text{ ms}^{-2}$

$a_1 = 2.98 \text{ ms}^{-2}$

$\omega_1 = 64.9^\circ$

(iii) If the wedge remains motionless, $\Sigma \vec{F}_{ext} = 0$

$3mg\cos 53^\circ \sin 53^\circ + S\sqrt{2}\sin 8^\circ - mg\sin 37^\circ \cos 37^\circ$
$- \eta (5mg + 3mg\cos^2 53^\circ + mg\sin^2 37^\circ + S\sqrt{2} \cos 8^\circ) = 0$

where η denotes the coefficient of friction.

Substituting values we obtain.

$$\eta = \frac{11.67}{81.9} \Rightarrow \underline{\eta = 0.142}$$

3.11 Problem 6

A series circuit consists of a capacitor C, an AC ammeter of neglegible impedance and a calibrated inductor whose inductance L can be changed by moving an iron core along the axis of its coil.

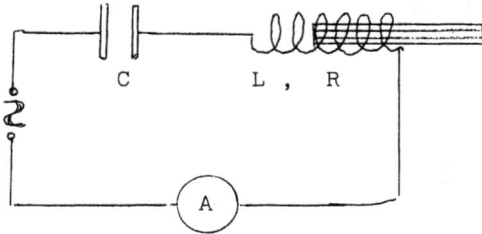

The circuit is fed from a power supply which provides a sinusoidal E.M.F of R.M.S value 2 V at constant frequency equal to $\frac{500}{2\pi}$ Hz

As the inductance of the inductor is changed it is observed that at some position the current through the ammeter becomes a maximum. Then, by moving the iron core it is observed that the current shown by th ammeter is reduced to $\frac{1}{\sqrt{2}}$ of its maximum value at two positions of the core. One of these corresponds to the inductor having an inductance $L_1 = 0.9$ H and the other to $L_2 = 1.1$ H.

Part (a)

Explain these observations and calculate the capacitance C of the capacitor and the resistance R of the inductor.

Part (b)

For each of the two values of L_1 and L_2 calculate

(i)   The impedance of the circuit

(ii) The current in the circuit

(iii) The phase difference between current and voltage.
      Draw a phasor diagram showing the voltages across
      each component of the circuit and explain whether
      current of voltage leads in each case.

3.12 Solution to Problem 6.

Part (a)

The circuit is a current resonant circuit and the
observations show that we are in a region close to
resonance.   The ammeter readings give R.M.S. values.

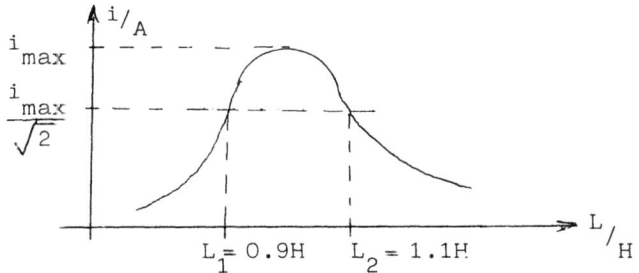

At resonance the current becomes a maximum i_{max}.   The
impedance Z takes its minimum value $Z_o = R$ where

$$Z_o = \frac{U_{RMS}}{I_{max}} = R$$

If Z_1, Z_2 denote the impedances at positions 1 and 2

$$Z_1 = \frac{U_{RMS}}{I_{max}/\sqrt{2}} \qquad\qquad Z_2 = \frac{U_{RMS}}{I_{max}/\sqrt{2}}$$

$$Z_1 = Z_2$$

$$\Rightarrow \quad \sqrt{R^2 + (\omega L_1 - \frac{1}{\omega C})^2} = \sqrt{R^2 + (\omega L_2 - \frac{1}{\omega C})^2}$$

$$(\omega L_1 - \frac{1}{\omega C})^2 = (\omega L_2 - \frac{1}{\omega C})^2 \quad \dots\dots(1)$$

if we take the positive signs of the square roots we obtain.

$$\omega L_1 - \frac{1}{\omega C} = \omega L_2 - \frac{1}{\omega C}$$

or $L_1 = L_2$ which is absurd since $L_1 \neq L_2$
(A mistake most sudents make).

However, since before resonance the voltage lags behind and after resonance it leads the current, we must take alternate signs i.e. equation 1 should yield

$$(\omega L_1 - \frac{1}{\omega C}) = -(\omega L_2 - \frac{1}{\omega C})$$

(The same result can be obtained by considering the phase angles between current and voltage in each case.)

$$\omega(L_1 + L_2) = \frac{2}{\omega C}$$

$$\Rightarrow \quad C = \frac{2}{\omega^2(L_1 + L_2)} \quad \Rightarrow \quad C = \frac{2}{500^2(1.1 + 0.9)}$$

$$\underline{C = 4\mu F}$$

Also at resonance

$$Z_o = \frac{U_{RMS}}{I_{max}} = R \quad \dots\dots(2)$$

At position 1 $\quad \dfrac{U_{RMS}}{I_{max}/\sqrt{2}} = Z_1 = \sqrt{R^2 + (\omega L_1 - \frac{1}{\omega C})^2} \quad \dots\dots(3)$

Equations (2) and (3) give

$$\sqrt{R^2 + (\omega L_1 - \frac{1}{\omega C})^2} = R\sqrt{2}$$

$$R^2 = (\omega L_1 - \frac{1}{\omega C})^2 \quad \text{or}$$

$$R^2 = (500 \times 0.9 - \frac{1}{4 \times 10^{-6} \times 500})^2$$

$$\Longrightarrow \quad \underline{R = 50 \, \Omega}$$

The negative value is ignored for the resistor.

Part (b)

(i)   The impedance Z_1 is

$$Z_1 = \sqrt{R^2 + (\omega L_1 - \frac{1}{\omega C})^2} \quad \Longrightarrow \quad Z_1 = 50\sqrt{2} \ \Omega$$

this is equal to Z2

(ii)   The current has a maximum value

$$I_{max} = \frac{U_{RMS}}{Z_o} \quad \Longrightarrow \quad I_{max} = \frac{U_{RMS}}{R}$$

$$I_{max} = \frac{2}{50} = 0.04 \ A \quad \text{and at each position the}$$

current is

$$I_1 = I_2 = \frac{0.04}{\sqrt{2}} = 0.0283 \ A$$

(iii) The phase difference between current and voltage at position 1 is given by

$$\tan \Phi = \frac{\omega L_1 - \frac{1}{\omega C}}{R} = \frac{500 \times 0.9 - (1/500 \times 4 \times 10^{-6})}{50}$$

$$\tan \Phi = -1 \qquad \Phi = -45^\circ$$

and at position 2 $\Phi = 45^\circ$

To construct the diagrams for each of the two

positions, the voltages across each component are calculated.
For position 1 we have:

Across the capacitor

$$V_c = I_1 \times \frac{1}{\omega C} \qquad\qquad V_c = 14.150 \text{ V}$$

The incuctance of the coil contributes with a potential
difference

$$V_L = I_1 \cdot \omega L_1 \qquad\qquad V_L = 12.735 \text{ volts.}$$

The resistance of the coil contributes with a potential
difference

$$V_R = I_1 R \qquad\qquad V_R = 1.415 \text{ volts}$$

now $V_c - V_L \doteq 14.150 - 12.735 = 1.415$ volts

Similar calculations give the voltages at position 2.

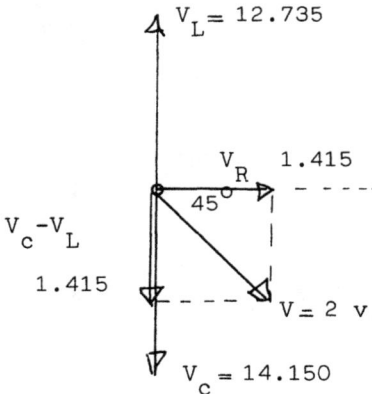

Position 1

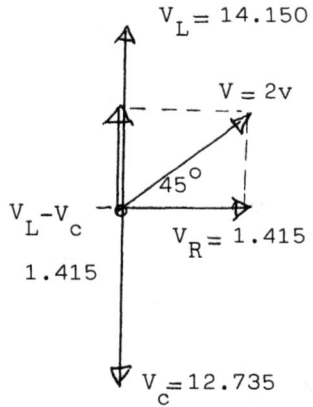

Position 2

FINLAND

Maija Ahtee
Jukka O. Mattila

THE PHYSICS COMPETITIONS IN FINLAND

1. The Finnish Educational System in General

In Finland the nine year comprehensive school is divided into a lower (primary) level, lasting six years and an upper
(lower secondary) level, lasting three years. The compulsory
comprehensive stage is followed by the upper secondary level
on which about 50% of the age group chooses the upper secondary school of three years and the other 50% some vocational
education with varying duration. As the school is started at
the age of seven, the pupils will normally be 19 years old
when finishing the upper secondary school.

The upper secondary school ends with the matriculation
examination at which the pupils are tested in four to six
subjects. Three of these are the mother tongue (mainly
Finnish), the second official language (mainly Swedish) and
the first foreign language (mainly English). The fourth
compulsory subject is either mathematics or a combined test
in social and natural sciences (including items from reli-
gion, history, biology, geography, physics, chemistry, psy-
chology and philosophy). As extra subjects one can choose
the other one of the two previously mentioned alternatives
or still another foreign language (usually either German or
French).

After the maturity examination the students are in principle
qualified to enter the universities. However, as there are
more students than the universities can accept, only a limited number can be taken in. In recent years it has become
more and more easy to enter the faculties prerequisiting
physics, because of the shortage of students having had
physics at the upper secondary.

General education in the meaning of language, literature and
social studies has traditionally been highly esteemed in the
Finnish educational system. Because of the very restricted
native language area, foreign languages cover quite a lot of
the pupil's daily schedule. As a consequence the total number of lessons for mathematics and science is low in international comparison.

In the Finnish comprehensive school mathematics is taught to all students. At the higher secondary there are two alternatives in mathematics which differ both in topics and in the amount of lessons. Physics and chemistry are taught at the comprehensive school on grades 7 to 9 for all pupils. At the upper secondary school only about 75 lessons of chemistry are taught as compulsory, while physics is not compulsory at all. Only about 25% of the pupils carry out the physics course which consists of about 300 lessons. Roughly 80 lessons of the course are devoted to mechanics, 30 to thermodynamics, 40 to waves and optics, 80 to electricity and magnetism, 40 to atomic and nuclear physics. The rest is devoted for repetition before the maturity examination. Contrasting the situation on the lower secondary, there are no lessons for pupil demonstrations on the upper secondary.

2. Structure of the National Physics Competitions

The main competition takes place on the 12th school year (last grade of the upper secondary), i.e. on the year preceding the entrance to the universities. However, a similar competition is held on the 11th school year to pick up the best pupils to be trained for the 12th year main competition.

11th school year competition. A nationwide competition is held in December. For practical reasons the competition is only theoretical, without any experimental part. The problems are prepared and distributed to all upper secondary schools by the physics competition board of the Association of Teachers of Mathematics, Physics and Chemistry, MAOL. At school the teacher decides whether the competition will be held for his/her pupils or not. The competition is thus completely voluntary. The best 30 pupils or so are picked up for extra training by correspondence, comprising usually of three letters with a set of problems each. The pupils are supposed to solve the problems and send their solutions to the board for inspection. The aim of the training is to prepare the pupils for the 12th year main competition.

12th school year competition. For practical reasons the 11th and 12th school year physics competitions are held at the schools simultaneously, usually at the beginning of December. The 12th year competition is in principle open for all pupils, i.e. also for the pupils of the 10th and 11th school year although it very seldom happens that they participate. The best 30 pupils are picked up for extra training by correspondence, again with three letters. Approximately 20 of the best are invited at the end of April to the national Selection Competition for the International Physics Olympiade.

Since 1988 the national Selection Competition for the International Physics Olympiade has been held at universities on circular basis. Lack of science students has activated the universities to seek publicity by hosting the competition and offering the students a possibility to get acquainted with the scientific work done at the university. For the same reason practically all of the mathematics, physics and chemistry faculties of the universities and technical universities allow free entrance for the 10 best of each of the corresponding national competitions.

The Selection Competition takes two days. On the first day the students solve both the theoretical and the experimental problems. Rest of the time is spent on excursions etc. The five best of the Selection Competition make up the team, in principle. In practice the same pupils are often on top of both the mathematics and physics competition and, unfortunately, often prefer the Mathematics Olympiade to the Physics Olympiade. Thus overlapping of the two international olympiades usually has harmful consequences for the physics team.

3. The System of Training of the Team to the IPhO

Because of the discrepancies between the national curriculum and the syllabus of the International Physics Olympiade as well as because of the total absence of pupil demonstrations at the Finnish upper secondary physics education, preparing of the team is essential. The gaps between the two syllabuses are tried to be covered by the letter training. For experimental training, the team spends the last week of May, in practice four days, at the Department of Physics at Helsinki University.
Both the selection and training processes are arranged on an unofficial basis by MAOL, the Association of Teachers of Mathematics, Physics and Chemistry. MAOL also takes care of the corresponding tasks for the mathematics and chemistry competitions.
The first national physics competition was held in Finland in 1969. Finland has participated the International Physics Olympiades since 1977.

THEORETICAL PROBLEMS

1. A homogeneous parallel plane straw with length l is originally at a horisontal position. It is supported from its centre so that it can rotate through its centre in a vertical plane. A spider falls with a vertical velocity v_0 and lands midway between the end of the straw and its centre. The mass of the spider is equal to the mass of the straw. Immediately after landing on the straw the spider starts to run along it so that the angular velocity of the straw stays constant. Determine the highest value of v_0 in case the spider reaches the end of the straw. We assume that the spider falls off the straw when the straw is at the vertical position. Draw the path along which the spider runs.

Solution

Denote the mass of the straw = the mass of the spider = m and the angular velocity of the straw = ω.

Then the moment of inertia of the system is $I = \frac{1}{12}ml^2 + mx^2$, where x is distance of the spider from the centre of the straw.

At the moment $t = 0$ when the spider lands on the straw $x = l/4$

According to the law of conservation the angular momentum before and after the spider lands on the straw is the same:

$$L = mv_0x = I\omega = \left(\frac{1}{12}ml^2 + mx^2\right)\omega .$$

From this equation the angular velocity of the straw can be solved:

$$\omega = \frac{12v_0}{7l}.$$

For rotational motion the twisting moment is given by

$$M = I\alpha = \frac{d(I\omega)}{dt} = \frac{dI}{dt}\omega$$

as ω is constant. From the figure it is seen that

$$M = mgx\cos\phi \text{ where } \phi = \omega t.$$

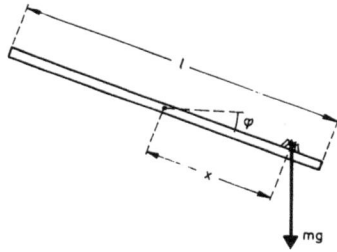

By combining we obtain the differential equation

$$mgx \cos(\omega t) = 2m\frac{dx}{dt}\omega .$$

From this the distance x can be solved

$$x = \frac{49l^2}{288\, v_0^2} \sin\left(\frac{12v_0 t}{7l}\right) + C.$$

The initial condition $t = 0$, $x = l/4$ gives
$C = l/4$.

The spider will fall down when $x = l/2$
and $\phi = \pi/2$. Its velocity is then

$$v_0 = \frac{7}{6}\sqrt{\frac{gl}{2}}.$$

In parameter form its path is described by

$$\phi = \sqrt{\frac{2g}{l}}t \ , \quad x = \frac{l}{4}(\sin\phi + 1).$$

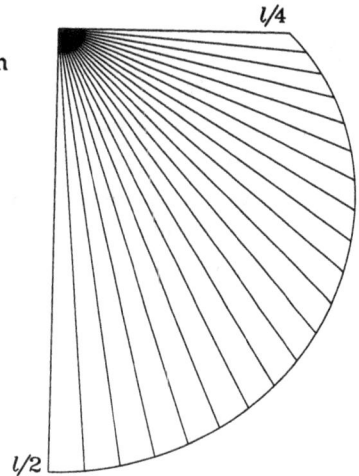

2. A liquid is boiled at different power levels using the system drawn in the figure.

Vapour is condensed and collected for 300 s at each power setting. In the table are the results of the measurement. Find the specific heat of vapourisation of the liquid. Discuss error sources and their elimination.

heater voltage (appr.)	energy	massflow
(V)	(kWh)	(g/300 s)
200	0.0237	26.0
200	0.0239	28.3
200	0.0240	27.9
180	0.0191	19.5
160	0.0150	14.2
140	0.0114	8.8
120	0.0083	3.9
120	0.0084	4.1
120	0.0085	4.2

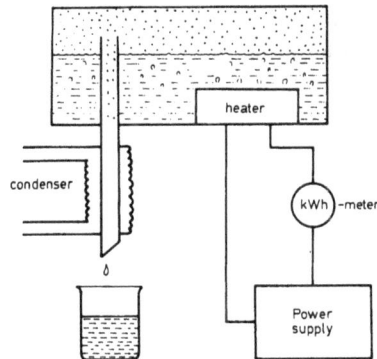

Solution

Denote L the specific latent heat , W_0 lost energy and m the mass of condensed liquid.

From the law of conservation of energy $W = mL + W_o$ the power developed in the heating system

$$P = \frac{dm}{dt} L + P_0$$

is obtained where dm/dt is the massflow. By plotting the power $P = W/t$ against the massflow a straight line is obtained. The slope gives the specific heat of vapourisation and lost power P_0 is obtained from the intersection with the horizontal axis. The result with the error is

$$L = (2420 + 150) \, J/g$$

and the lost power $P_0 = 67$ W.

3. Around an iron core with a shape of a toroid is wound a primary coil through which flows an electric current I_1 which is changed by steps as shown in the table. A ballistic galvanometer G measures the induced charges Q_2 when the primary current is changed. The number of turns in windings in the primary coil is 250, and 50 in the secondary coil. The resistance of the secondary coil is 10.0 Ω, and the resistance of the ballistic galvanometer is 40.0 Ω. Draw the hysteresis curve (H, B) (magnetic field strength, magnetic flux density) according to the measurement when the length of the toroid is 50.0 cm and the cross-section 10.0 cm^2.

A toroid has the same form as the inner tyre of a car wheel. When the wire is wound evenly around the toroid, the magnetic field is confined inside the ring and the magnetic flux density has the form $B = \mu NI/l$, where N is the number of turns in winding of the wire which carries the current I; l is the length of the toroid and μ the permeability of the material.

I_1(A)		$Q_2 \, (\times 10^{-5} \, C)$
2.0	---> 1.0	- 8.0
1.0	---> 0.50	- 17.0
0.50	---> 0	- 42.0
0	---> -0.50	-200.0
-0.50	---> -1.0	- 39.0
-1.0	---> -2.0	- 14.0
-2.0	---> -1.0	+ 8.0
-1.0	---> -0.50	+ 17.0
-0.50	---> 0	+ 42.0
0	---> 0.50	+ 200.0
0.50	---> 1.0	+ 39.0
1.0	---> 2.0	+ 14.0

Solution

When the primary current is changed the potential difference is induced in the secondary coil

$$e = -N_2 \frac{d\Phi}{dt} = -N_2 A \frac{dB}{dt},$$

where Φ is the magnetic flux through the toroid and A its cross-section.

The induced current in the secondary coil is thus

$$I_2 = \frac{e}{R} = -\frac{N_2 A}{R}\frac{dB}{dt},$$

where R is the resistance of the secondary coil and the ballistic galvanometer together. The quantity of electricity through the galvanometer is then

$$Q_2 = \int_{I_1}^{I_2} I_2 dt = -\frac{N_2 A}{R} \int_{B_1}^{B_2} dB = -\frac{N_2 A}{R} \Delta B.$$

The changes in the magnetic flux density are thus given by the different values of Q_2. The corresponding values of the magnetic flux density B are obtained by adding up the ΔB values. The magnetic field is given by $H = N_1 I_1 / l$. By plotting the corresponding values of H and B the hysteresis curve can be drawn.

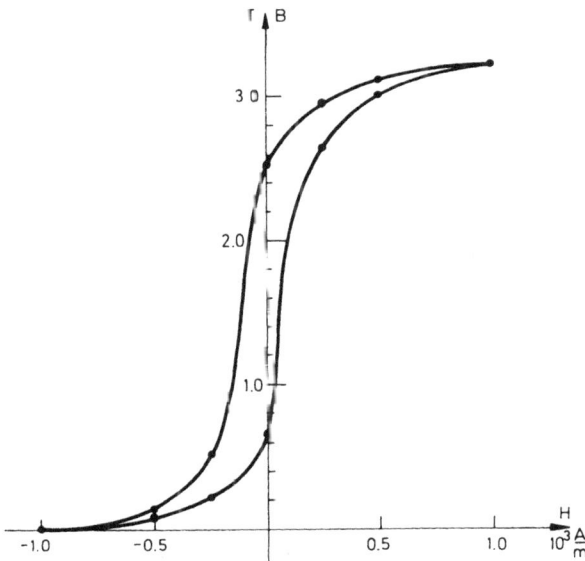

4. A uniform homogen board rests on two horizontal fixed rods A and B at an angle Θ. The distance between the rods is d and the coefficient of static friction between the rods and the board is μ. Determine the distance of the centre of gravity of the board from the rod A when the board is not to slip.

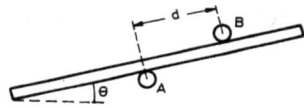

Solution

The centre of gravity P of the board has to lie to left from the rod A. Denote x (= PA) the shortest distance for the board not to slip.

The forces acting at the board are shown in Figure: $F\mu$ the force of static friction, N the normal force and G the force of gravity.

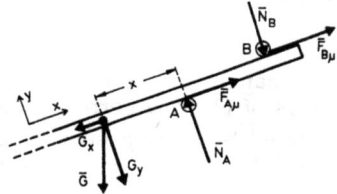

The first condition for equilibrium in the horizontal and vertical direction is:

$x: F_{B\mu} - F_{A\mu} - G_x = 0$

$y: N_A - N_B - G_y = 0$

where $F_{B\mu} = \mu N_B$ and $F_{A\mu} = \mu N_A$

$G_x = G \sin\Theta$ and $G_y = G \cos\Theta$

Let us choose the position of the rod A as the pivot point. Then according to the second condition for equilibrium the sum of the torques must be zero:

$G_y x - N_B d = 0$

By solving from these three equations we obtain

$$x \geq \frac{d}{2\mu}(\tan\theta - \mu).$$

5. A comet approaches the Sun far out in the space with the velocity v_o and the perpendicular distance d from the Sun. Determine the maximum velocity and the shortest distance of the comet in its orbit around the Sun.

Solution

The equation of motion for the comet
$$\overline{F} = m\overline{a} = -\gamma \frac{mM}{r^2}\overline{e}_r,$$

where $\overline{e}_r$ is the unit vector from the Sun to the comet along the distance $\overline{r}$.

Initially, when the comet is far from the Sun it has only kinetic energy $E_{kin} = \frac{1}{2}mv_0^2$, (as $r_0 \approx \infty$ its potential energy is zero) and its angular momentum is
$$\overline{L} = m\left|\overline{r}_0 \times \overline{v}_0\right| = mv_0 d.$$

When it is nearest to the Sun its maximum velocity v_{max} and shortest distance r_{min} are perpendicular so that
$$\overline{L} = m\left|\overline{r}_{min} \times \overline{v}_{max}\right| = mr_{min}v_{max}.$$
and
$$E_{tot} = \frac{1}{2}mv_{max}^2 - \gamma\frac{mM}{r_{min}}.$$

As there are no external forces affecting in the system both the law of conservation of energy and of angular momentum hold:
$$\frac{1}{2}mv_0^2 = \frac{1}{2}mv_{max}^2 - \gamma\frac{mM}{r_{min}}$$

$$mv_0 d = mr_{min}v_{max}.$$

From these equations the results are obtained
$$v_{max} = \frac{\gamma M + \sqrt{\gamma^2 M^2 + v_0^4 d^2}}{v_0 d} \quad \text{and} \quad r_{min} = \frac{\sqrt{\gamma^2 M^2 + v_0^4 d^2} - \gamma M}{v_0^2}.$$

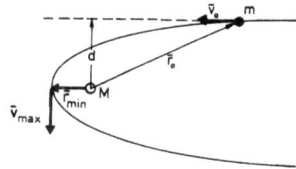

6. In Young's experiment 600 nm light falls perpendicularly on two closely spaced narrow slits. The distance of the screen from the slits is 1.00 m and the distance of the central peak from the tenth bright line is 300 mm.

a) Determine the slit width.

b) When thin transparent 0.02 mm thick film is placed on the other slit the central peak is displaced 30 mm. Determine the index of refraction of the film.

Solution

a) The tenth bright line will occur on the screen when

$d \sin \Theta = n\lambda,$ where $n = 10$.

From Figure we obtain

$\tan \theta = \frac{a}{l}.$

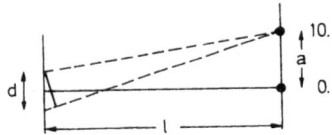

This corresponds a very small angle, so that we can take $\sin \Theta \approx \tan \Theta$ (a reasonable approximation when $\Theta < 10°$).

Thus

$$d = \frac{n\lambda}{\sin \Theta} \approx \frac{n\lambda l}{a} = 0.20 \text{ mm.}$$

b) We have to determine the path difference of light travelling in air and in the film. Denote the thickness of the film with s. The time needed in air for this distance is

$t_a = \frac{s}{c}.$

The extra path length is thus in air

$\Delta = c \left(t_f - t_a\right) = s \left(\frac{c}{c_f} - 1\right) = s \left(n - 1\right),$

where n is the index of refraction of the film.

According to part a) the displacement of the central peak corresponds the distance of the tenth bright line: $\Delta = 10\lambda$. Thus

$n = 1 + \frac{\Delta}{s},$

which gives for the index of refraction $n = 1.3$.

EXPERIMENTAL PROBLEMS

1. Fill a glass with water, cover it with a card and turn then upside down. The water stays in. What happens if the glass is only partly filled with water? Try it. Explain it.

Solution

When the glass is only partly filled the paper puffs out slightly or a small amount of water drips out. The air pressure inside the glass decreases from p_0 to p.

Denote the height of the glass with l, height of water with h and the change in the water height with Δh.

Let us assume that the ideal gas law holds: $pV = $ constant. Then

$$p_0 A (l - h) = pA(l - h + \Delta h)$$

If the paper sticks on the glass, then

$$p = p_0 - \rho g h, \text{ where } \rho g h \ll p_0.$$

The change in the water height is

$$\Delta h \approx \frac{\rho g h \, (l - h)}{p_0}.$$

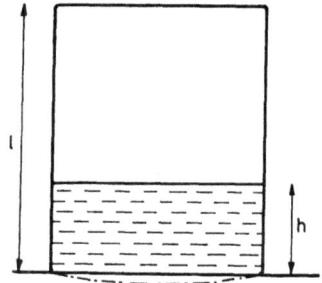

2. Wind a string around a roller so that it comes out on the lower side (see Figure). Pull from the string at different angles. Explain.

Solution

The forces acting at the roller are: the force of gravity G, the normal force N and the static friction F_μ, where $F_\mu = \mu N$.

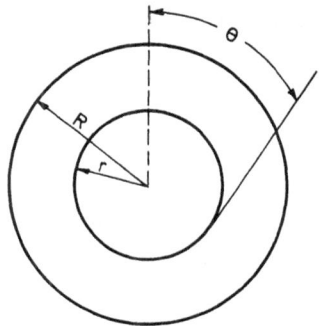

When the roller stays at rest the first
condition of equilibrium gives

x: $F \sin \Theta - F_\mu = 0$ (1)

y: $N + F \cos \Theta - G = 0$ (2)

and the second condition of equilibrium

$F r - F_\mu R = 0$ (3)

Equations (1) and (3) determine the critical angle

$\sin \Theta_0 = r/R$

From Eq. (2) it is obtained

$F_\mu = \mu(G - F \cos \Theta)$

The roller

1. rolls to left when $\Theta < \Theta_0$
2. rolls to right when $\Theta > \Theta_0$
3. stays at rest when $\Theta = \Theta_0$ or slides when $F \sin \Theta_0 > F\mu$.

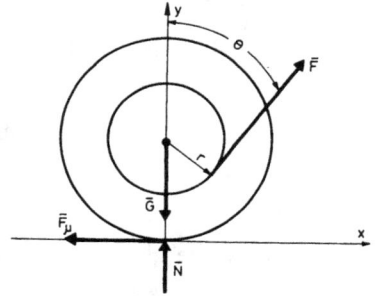

3. Two plane mirrors form an angle α as
shown in Figure. Find the number of
pictures formed in the mirrors.

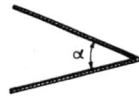

Solution

The case when $\alpha = 90°$ is shown here. Three images are obtained.

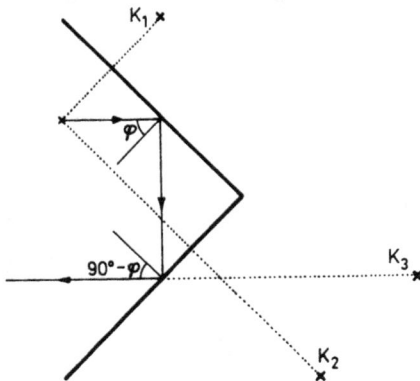

4. Determine the elastic properties of a string as a function of the stretching force.

In evaluating the problem attention is paid to the measuring method and the accuracy of the result. Write a report from your work giving a detailed measuring protocol as a supplement.

Only given equipments are allowed.

GERMAN D. R.

THE NATIONAL PHYSICS COMPETITION OF
THE GERMAN DEMOCRATIC REPUBLIC AND
THE PROCEDURE FOR SELECTING THE TEAM
FOR THE INTERNATIONAL PHYSICS OLYMPIADS

Rudolf Gau and Christoph Schick

Pädagogische Hochschule "Liselotte Herrmann" Güstrow

Sektion Mathematik/Physik

Goldberger Str. 12

2600 - Güstrow / GDR

1. THE NATIONAL PHYSICS COMPETITION OF THE GDR

The national competitions in mathematics, physics, chemistry and informatics are arranged in accordance with the well-planned school policy of the GDR, which is aimed at the all-round development of the individuality of all pupils on the basis of a profounded uniform general education.

Since 1972 the central physics competitions of the GDR for physically interested and particularly gifted pupils have yearly been held in the Pedagogical University "Liselotte Herrmann" Güstrow Physically interested and particularly gifted pupils, who have acquired knowledge and abilities in the physics instruction of the general polytechnical schools, in special schools with orientation to mathematics, natural sciences and technology, in courses and academies of pupils, can measure their intellectual forces and check their abilities of solving physical problems. The national physics competition is carried out in two stages.

The first stage is a qualification examiniation (theoretical problems only). It is organized in three categories with different problems.

* For pupils of the 9th form,
* for pupils of the 10th form,
* for pupils of the 11th and 12th forms.

The problems and the evaluation scheme are prepared by a

commission of the Pedagogical University "Liselotte Herrmann" Güstrow. The organization of this first stage is
realized according to the order of the regional educational
boards. At the beginning of each school year the educational
authorities notify all schools about the organization of the
central physics competition. The examinations of the first
stage are organized on the same day in September in all
regions of the GDR. All physically interested and particularly gifted pupils are invited to these examinations.
Many of the participants are pupils from the special schools
with orientation to mathematics, natural sciences and technology.

The solutions and evaluations of the best pupils in every
district are then submitted to the selection commission
responsible for the further procedure. After the coordination of the evaluation, the best pupils of each category are
invited to the second stage of the competition.

This second stage is the finale of the national physics competition and takes place every year in February during the
schools' winter holidays in the Pedagogical University
"Liselotte Herrmann" Güstrow. It is divided into a theoretical and an experimental examination, being different for the
two categories (9/10 and 11/12 forms).

On the first day of the competition the pupils have to solve
four theoretical problems and on the second day one experimental problem. The working time in each case runs to four
or five hours. Compared with the standard of the International Physics Olympiads the requirements in the national competitions are lower in the category 11/12 and much lower in
the category 9/10.

The same principles as in the IPhO are used for the organization of the competition, for the evaluation scheme and for
the determination of the prizes for each category.

The winner of the awards of the category 9/10 acquire further theoretical and experimental knowledge outside their

school instruction in special courses during the schools' summer holidays. Theses courses are organized by one of the special schools with orientation to matematics, natural sciences and technology. The best pupils of the category 11/12 are invited to a preparatory course for the International Physics Olympiads.

The problems and results of the national competition in physics have always been published in the journal "Physik in der Schule".

2. THE SELECTION PROCEDURE FOR THE TEAM TO THE IPHO

The selection procedure and the preparetion of the team for the International Physics Olympiads are organized in four stages, including the two stages of the national competition. After the first stage of the national competition in September the best pupils in the category 11/12 are invited to a preparatory course. This course is held in the Pedagogical University in Güstrow during the holidays in October and is the second stage in the selection procedure. Since the syllabus of the secondary schools of the GDR is quite different from that of the IPhO, it is necessary to instruct the candidates with completely new facts. These are special parts of classical mechanics, thermodynamics and wave optics. To quantum physics, relativity and physics of matter only a short introduction is given in school instruction. Therefore the main concern of the preparatory courses is to make the candidates acquainted with the basic relations in modern physics.

Within two weeks the pupils take part in lectures and exercises, discuss problems and carry out experiments. The lectures and exercises are given by professors of the university, by teachers from special schools and by previous participants of the IPhO. Theoretical and experimental examinations are used to check the knowledge of the course participants.

All participants of the preparation course are invited to the national competition too, the third stage in the selection procedure. On the basis of the results obtained in the national competition the best 10 pupils in the category 11/12 are selected and invited to the second preparatory course.

This 4th stage in the preparation of the team is organized in the spring holidays in the same way as the 1st course.

In the end of the 2nd preparatory course the five members of the team and another reserve student are selected according to their results obtained in the examinations of the national competition and the preparatory courses.

Beside this selection procedure for the team to the IPhO all candidates participate in a correspondence, with colleagues of the Güstrow Physical Department who make them familiar with problems and questions of selected fields of physics. These letters support the acquiring of determined demands, corresponding to the statute of the IPhO, stimulate the independent elaboration of special fields in physics and contribute to develop the pupils abilities of solving physical problems.

The five selected team members and the reserve pupil take part in a training course two or three weeks before leaving for the IPhO. This course is intended to give them practical training in solving theoretical and experimental problems with a high level of requirement. They do experimental tests from former international olympiads and follow lectures about error estimation and experimental set-ups.

3. THEORETICAL AND EXPERIMENTAL PROBLEMS

Problem 1:

A plate capicitor with aer as dielectric is arranged horizontally. The lower plate is fixed and the other connected
with a perpendicular spring. The area of each plane is A.
In the steady position the distance between the plates is
d_o and the frequency for the oscillating plate is ω_o.
When the capicitor is connected with an electric source with
the voltage U a new equilibrium distance d_1 appears.

a) Determine the spring constant k!

b) What is the maximum voltage for a given k, in which an
   equilibrium distance is possible?

c) What is the frequency of the oscillating system around
   the equilibrium distance d_1 ? Which are the possible
   values for a stable equilibrium distance d_1?

Remark: The amplitude of the oscillation is much more lower
        than the distance of the plates.

Solution Nr.1:

a) If the voltage U is layed out on the capacitor, the
system is oscillating by the force

$$F = k \cdot (d - d_o) + \frac{1}{2} Q \cdot E = -k (d_o - d) + \frac{1}{\varepsilon_o} A \cdot \frac{U^2}{d^2} \qquad (1)$$

The conditions for the equilibrium are d = d_1 and F = 0.
For the spring constant k we get from (1)

$$k = \frac{\varepsilon_o \cdot A \cdot U^2}{2 (d_o - d_1) d_1^2} \qquad (2)$$

b) For the maximum voltage the condition

$$\frac{d U^2}{d d_1} = 0$$

must be satisfied.

From equation (2) we get for the equilibrium distance

$$d_1 = \frac{2}{3} d_0 \tag{3}$$

and therfore for the maximum voltage

$$U \leq U_{max} = \sqrt{\frac{k}{\varepsilon_0 \cdot A}} \left(\frac{2}{3} d_0\right)^{3/2} \tag{4}$$

c) For the oscillation around the equilibrium distance d_1 we get from (1) with $d = d_1 - x$

$$F = - k \left(d_0 - d_1 + x\right) + \frac{1}{2} \varepsilon_0 \cdot A \cdot \frac{U^2}{d_1} \cdot \frac{1}{\left(1 - \frac{x}{d_1}\right)^2} \tag{5}$$

For small amplitudes the last term in the brackets can be expanded in a power series of x/d_1. In the linear order we get from (5)

$$F = - \left(\frac{k - \varepsilon_0 \cdot A \cdot U^2}{d_1^3}\right) \cdot x = - k' \cdot x \tag{6}$$

The frequency for the oscillation is given by $\omega = \sqrt{\frac{k'}{m}}$
From (2) and (6) we get

$$\omega = \sqrt{\frac{k}{m} \left(\frac{3d_1 - 2d_0}{d_1}\right)}$$

The condition for a stable equilibrium is given by

$$3d_1 - 2d_0 \geq 0$$

Problem 2:

Electrons are accelerated by a voltage U in an evacuted tube. The electrons are going in a magnetic field B after the acceleration, which has an orientation perpendicular to the velocity of the particles. The trajectory of the electrons is a circle with the radius R. The specific charge e/m of the electrons can be determined by the voltage U, the strength of the magnetic field B and the radius R of the trajectory.

a) Determine the specific charge of the electrons, following the classical electrodynamics!

b) What is the general relation for the determination of the specific charge of the electrons? Compare and discuss the difference between the results!

Solution Nr.2:

a) The conservation of energy in the classical form is given by the relation

$$\frac{m}{2} v^2 = e \cdot U \tag{1}$$

The trajectory of the particles is determined by the equilibrium of the radil-force and the Lorenz-force

$$\frac{m \cdot v^2}{R} = e \cdot v \cdot B \tag{2}$$

From this basic relation we get for the specific charge

$$\frac{e}{m} = \frac{2 \cdot U}{R^2 \cdot B^2} \tag{3}$$

b) The relativistic form of the equations (1) and (2) must be used for the calculation of e/m in the general case. The conservation of energy is given by

$$\frac{m_0 c^2}{\sqrt{1 - \beta^2}} - m_0 c^2 = e \cdot U \qquad \left(\beta = \frac{v}{c}\right) \tag{4}$$

and from the equilibrium of the forces we get

$$e \cdot v \cdot B = \frac{m_0 \cdot v^2}{R \sqrt{1 - \beta^2}} \tag{5}$$

The specific charge is given by (6), determined by the relations (4) and (5).

$$\frac{e}{m} = \frac{2 \cdot U}{B^2 \cdot R^2 - \frac{U^2}{c^2}} \tag{6}$$

The comparison of (3) with (6) proves that the relativistic relation (6) can be used only for $U/c \lesssim 1$.

The voltage which creates for example a difference from 1 % in the different calculations of the specific charge can be calculated as follows. From (4) we get the relation

$$m = m_o \cdot \left(1 + \frac{e \cdot U}{m_o \cdot c^2}\right) \qquad \text{with} \qquad m = \frac{m_o}{\sqrt{1 - \beta^2}}$$

For the relation $m = 1.01 \, m_o$ follows $U = 5.11$ kV.

Problem 3:

The law of STEFAN-BOLTZMANN gives a relation between the emissions u of a black body as a function of his temperature T.

$$U = \sigma \cdot T^4 \qquad \qquad \sigma = 5.67 \cdot 10^{-8} \quad W \cdot K^{-4} \cdot m^2$$

The current of the energy density on the surface of the earth per unit of time is mesured by experiments. It is
$$S = 1.36 \ast 10^3 \ W/m^2$$
Calculate the middle temperatures on the surface of the sun and the earth! Assume that the sun and the earth can be considered as black bodies and the total current of energy density on the surface of the earth is created only by the radiation on the sun.

The radius of the sun is $\qquad R_S = 7 \ast 10^8$ m.
The middle distance sun-earth is $R_{SE} = 1.5 \ast 10^{11}$ m.

Solution Nr.3:

The energy N emited by the sun per unit of time is in accordance with the law of STEFAN-BOLTZMANN

$$N = 4\pi \cdot R_S^2 \cdot \sigma \cdot T^4$$

(1)

This energy is distributed on the surface of a sphere. The radius of this sphere in the distance sun-earth is given by

R_{SE}. That means, the energy is approximately given by

$$N = 4\pi \cdot R_{SE}^2 \cdot S$$

(2)

From (1) and (2) we get for the temperature on the surface of the sun

$$T_S = \sqrt{\frac{R_{SE}^2 \cdot S}{R_S^2 \cdot \sigma}} = 5.8 \cdot 10^3 \; K.$$

The middle temperatur of the earth can be calculated by a similar consideration. The energy absorbed by the earth per unit of time is

$$N = 4\pi \cdot R_E^2 \cdot S$$

(3)

If we assume, that the earth is a black body, the same part of energy is emitted by the earth. That means

$$N = 4\pi \; R_E^2 \cdot \sigma \cdot T_E^4$$

(4)

From (3) and (4) we get for the temperature on the surface of the earth

$$T_E = \sqrt{\frac{S}{4 \cdot \sigma}} = 280 \; K.$$

Problem 4:

A sphere with the mass M and the radius R is located on a horizontal table. The sphere gets the velocity v_c by a horizontal linear momentum in direction of the center of mass. The coefficient of friction between the table and the sphere is μ.

a) Determine the distance where the initial glide motion of the sphere turns into a rolling motion!

b) Where the horizontal linear momentum must be given, that the initial motion of the sphere is a rolling motion?

Solution Nr.4:

The motion of the sphere caused by the horizontal linear momentum is a superposition of a linear motion and a rotary motion.

The initial angular momentum is zero because the linear momentum is given in direction of the center of mass. The friction is the reason for a torque around the center of mass. This is given by

$$M_S = \mu \cdot M \cdot g \cdot R$$

(1)

The angular acceleration $\dot{\omega}_o$ can be calculated from (1) with the moment of inertia for a sphere J_S . It is

$$\dot{\omega}_c = \frac{5}{2} \mu \cdot \frac{g}{R}$$

(2)

From this relation we get the angular velocity for the rotary motion of a sphere by

$$\omega_c = \frac{5}{2} \cdot \mu \cdot \frac{g}{R} \cdot t$$

(3)

The linear motion of the sphere is determined by the law of NEWTON in the form

$$M \cdot a = -\mu \cdot g \cdot M$$

(4)

(a - is the linear acceleration)

We get the linear velocity v_S and the distance as a funtion of the time from (4) by simple integrations.

$$V_S = V_0 - \mu \cdot g \cdot t \quad , \qquad\qquad S = V_0 \cdot t - \frac{\mu}{2} g \cdot t^2$$

(5)

If the condition

$$V_S = \omega_0 \cdot R$$

(6)

is satisfied, the initial glide motion turns into a rolling motion. This condition allows to calculated the instant of time for this process. We get from (6) with (3) and (5) for the time

$$t = \frac{2}{7} \frac{V_0}{\mu \cdot g}$$

and finally for the distance, where the rolling motion begins

$$S = \frac{12}{49} \frac{V_0^2}{\mu \cdot g}$$

b) The horizontal linear momentum must be given in a distance R+h from the table, if the initial motion of the sphere is a rolling motion. This linear momentum creates an angular momentum L:

$$L = M \cdot v_o \cdot h = J_s \cdot \omega \tag{7}$$

The angular velocity is determined from this relation by

$$\omega = \frac{5}{2} \frac{v_e \cdot h}{R^2} \tag{8}$$

and we get as condition for a initial rolling motion from (6) and (8)

$$h = \frac{2}{5} R.$$

That means, the horizontal linear momentum must be given in a distance h above the center of mass of the sphere.

Problem 5:

The focal length f of a lens is correlated with the frequency of the light. The reciprocal constringence is used for the discription from this correlation, which is given by

$$\frac{1}{\nu} = \frac{n_D - 1}{n_H - n_c}$$

n_D – refractive index of the D-line

n_H – refractive index of the H-line

n_c – refractive index of the C-line

The principle position of these lines in a spectrum is shown in the figure

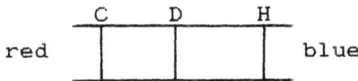

What are the conditions for the focal lengths and the reciprocal constrigence for a lens system constructed by two lenses which satisfy the relation

$$f_H - f_c = 0$$

The combination shall be a convergent lens. The refractive

indexes of the lenses are n_1 and n_2 and the focal lengths are f_1 and f_2. These values are the values of the D-lines.

Remarks: - The relation for the focal length of the system is given by

$$\frac{1}{f} = \frac{1}{f_1} + \frac{1}{f_2}$$

- For each lens the following relation can be used

$$\frac{1}{r_1} + \frac{1}{r_2} = R \qquad (r_{1,2} - \text{radius of curvatures})$$

Solution Nr.5:

For the focal lengths of the two lenses we get the relations

$$\frac{1}{f_{D1}} = (n_{D1}-1)\cdot R_1 \qquad\qquad \frac{1}{f_{D2}} = (n_{D2}-1)\cdot R_2$$

$$\frac{1}{f_{C1}} = (n_{C1}-1)\cdot R_1 \qquad\qquad \frac{1}{f_{C2}} = (n_{C2}-1)\cdot R_2 \qquad (1)$$

$$\frac{1}{f_{H1}} = (n_{H1}-1)\cdot R_1 \qquad\qquad \frac{1}{f_{H2}} = (n_{H2}-1)\cdot R_2$$

These relations can be transformed into

$$\frac{1}{f_{H1}} - \frac{1}{f_{C1}} = (n_{H1} - n_{C1})\, R_1 \quad , \quad \frac{1}{f_{H2}} - \frac{1}{f_{C2}} = (n_{H2} - n_{C2})\, R_2 \quad (2)$$

From (1) and (2) we get a relation for the reciprocal constringence for each lens

$$\frac{\dfrac{1}{f_{D1}}}{\dfrac{1}{f_{H1}} - \dfrac{1}{f_{C1}}} = \frac{(n_{D1} - 1)}{(n_{H1} - n_{C1})} = \frac{1}{\nu_1}$$

and for the focal lengths

$$\frac{1}{f_{H1}} - \frac{1}{f_{C1}} = \frac{\nu_1}{f_{D1}} \quad , \quad \frac{1}{f_{H2}} - \frac{1}{f_{C2}} = \frac{\nu_2}{f_{D2}} \qquad (3)$$

Whith the relation given in the problem, we get from (3)

$$\frac{\nu_1}{f_1} = - \frac{\nu_2}{f_2} \; . \qquad\qquad\qquad (4)$$

That means, one lens of the system must be a convergent lens and the other a divergent lens.

Presupposing the combination being a convergent lens, we get
from the condition

$$\frac{1}{f_1} + \frac{1}{f_2} > 0$$

$v_2 < v_1$ if f_2 – is the focal length of the convergent lens
ou $v_1 < v_2$ if f_1 – is the focal length of the convergent lens.
That means, the reciprocal constringence of the convergent
lens must be lower than that of the divergent lens.

Problem 6:

1. Construct a set-up to investigate the FRANK-HERTZ-charac-
teristic $I_a = f(U)$ for the given thyratron filled with rare
gas!

2. Take the characteristic from the oscillograph screen
(scaled drawing). Determine the activation energy of the
rare gas, the energy of the emitted photons and the frequency and wavelength of the radiation!

3. Discuss the results and estimate the errors!

Instrumentation:

1 oscillograph,

1 power suply (a.c. 20 V, 6.3 V),

1 thyratron (S 0.5/0.11 V, RFT Röhrenwerk Berlin),

1 resistor 100 Ω ,

1 resistor 2 MΩ,

1 adjustable resistor 100 Ω ,

1 multimeter,

12 connecting wires.

Hint:

You have to use a smaller heating voltage than 6.3 V (use
the adjustable resistor)!

The first maximum may be wrong due to contact potentials.
Use the first and second maximum! It is not necessary to use
a small voltage between anode and cathode! Determine the
scaling factor for the x-axis by yourself! Use the 100 Ω
resistor as a protection resistor in the anode circuit!

There is a phase shift between movement to the rigth and the return movement of the cathode ray. Use only one of them!
Constant: $h = 6.63 * 10^{-34}$ Ws^2.
(Experimental problem of the XVI. National Physics Competition of the GDR; by H. Schewelies)

Solution Nr.6:
1. Wiring diagram   (4 points)

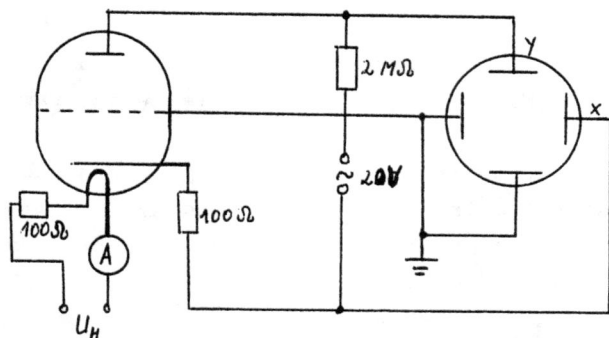

2. I (U)-characteristic (diagram)   (4 points)
- Determination of the scaling factor for the   (1 point)
  x-axis.
- Determination of the activation energy   (3 points)

$$\Delta E \approx 15 \cdot 10^{-19} \text{ Ws}$$

- Energy of the photons   (1 point)

$$E_{photon} = \Delta E$$

- Frequency and wavelength of the radiation   (2 points)

$$\Delta E = h \cdot f \qquad , \qquad f = 2.3 \cdot 10^{13} \text{ Hz}$$

$$\lambda = c/f \qquad \lambda = 129 \text{ nm}$$

3. Error estimation   (2 points)

$$\Delta E = (15 \pm 2)\ 10^{-19}\ Ws \qquad f = (2.3 \pm 0,2)\ 10^{13}\ Hz$$
$$\lambda = (129 \pm 12)\ nm$$

4. Discussion

- The droping in the I_a (U)-characteristic is a result of inelastic interaction between the electrons and atoms in case the kinetic energy of the electrons equals the activation energy of the atoms. (1 point)
- At higher voltages it will happen that the electrons reach ones move the activation energy after a inelastic interaction (second droping). (1 point)
- After 10^{-8} s the electrons go down to the ground state. The energy is irradiated as a photon with $\lambda \approx$ 130 nm. That is UV-radiation and not visible. (1 point)

Total: 20 points

Problem 7:

1. Determine the wiring diagram and the values of the components for the unknown four-pole inside the given black box!
2. Determine the errors!

Instrumentation:

1 black box,

1 power supply (2 ... 20 V, d.c., a.c. 50 Hz),

2 multimeter,

1 adjustable resistor ($\approx 100\,\Omega$),

7 connector wires.

Hint:

In every section of the wiring there is no more than one passive component! There are no capacitors and coils. Use only voltage less than 10 V!

(Experimental problem of the XVII. National Physics Competition of the GDR, 9/10 forms, by H. Schwelies)

Solution Nr.7:

1. Wiring diagram for the measurments. (1 point)
2. Measure the voltage between the different points.

Result: There is no power source inside the box. (1 point)
3. Check up the I(U)-characteristics between the different
   points. Change the polarity!
   Result: There is a on stage region and after changing the
           polarity a cut off region. The I(U)-characte-
           ristic is a straight line with a point of inter-
           section near 0.7 V.
There is a combination of a diode and a resistor between 1-3
and 2-4. The resistor results the straight line of the characteristic. From the slop of the characteristic you can
calculate the value of the resistors because the resistor of
the diode is negligible.

$$R_{13} = 156 \; \Omega \; ,$$ (2 points)

$$R_{24} = 373 \; \Omega \; .$$ (2 points)

Between 1-4 there is only a combination of resistors (one
straigth line for both polarities).

$$R_{14} = 658 \; \Omega \quad \text{that means } R_{14} > R_{13} + R_{24} \; .$$

There must be a third resistor. (2 points)
Between 2-3 you cant see any current for voltages in the
range −10 to +10 V. There may be two antiparallel diodes
(serial connected) in combination with a unknown resistor.

(2 points)

The resulting wiring diagram may be the following:

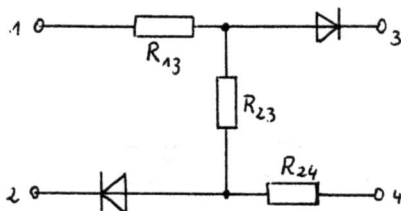

(2 points)

$$R_{23} = R_{14} - R_{13} - R_{24} = 129 \; \Omega$$ (2 points)

Revision the values of R_{12} and R_{34}:
The wiring diagram should be rigth if the following rela-

tions are satisfied: $R_{12} = R_{13} + R_{23}$ and $R_{34} = R_{23} + R_{24}$.
Determining R_{12} and R_{34} from the corresponding $I(U)$-characteristics between 1-2 and 3-4. The relations are satisfied.

(2 points)

the conclusion that the wiring diagram is the (2 points)
right.

4. Error-estimation for the characteristics and the resistors. (2 points)

The black box contens was: 2 diodes SY 305,

2 resistors 150 Ω + 5 %,

1 resistor 360 Ω + 5 %.

Total: (20 points)

Problem 8:

1. Determine the wiring diagram and the values of the components for the unknown four-pole inside the given "black box"!

2. Determine the errors!

Instrumentation:

1 black box,

1 power supply (2 ... 20 V; d.c. a.c. 50 Hz),

2 multimeter,

1 adjustable resistor,

7 connecting wires.

Hint:

In every section of the wiring there is no more than one passive component!

(Experimental problem of the XVII. National Physics Competition of the GDR, 11/12 forms, by H. Schewelies)

Solution Nr 8:

1. Wiring diagram for the measurments of voltage and current. (1 point)

2. Measuring the voltage between the different points.
Results: Between 3-4 and 3-2 there is a voltage of 1.5 V.

```
(-) 3----4 (+)
(-) 3----2 (+)                                    (1 point)
```

3. To answer the question whether there is a combination of
   a battery and a resistor or not you have to determine the
   U(I)-characteristic with the help of the   adjustable resistor to get the open-circuit voltage and the short-cir-
   cuit current between points 3-4 and 3-2.

   Result: In both cases there are the same   values   for U_Q
           and I_K. R = 1490 Ω .

Normally the battery resistance is in the order of less than
1 Ω . That means there is a combination of a resistor and   a
battery.

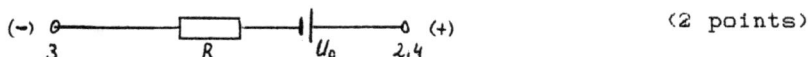

```
(-) o────────[    ]────┤├────o (+)          (2 points)
  3          R        U₀   2.4
```

The structure of the resulting resistor R   is   yet   unknown.
                                                      (1 point)

4. Because there are the same values for U_0 and I_K between
   3-4 and 3-2 you have to prove if there is a   direct   connection between 2-4. With the   help of   the   adjustable
   resistor you can prove it.

   Result: There is a direct connection between 2-4.

                                                   (1 point)

5. You have to prove the   behaviour   between   1-3   and   1-2.
Using an ammeter and d.c. voltage you can   observe   a   pulse
current. There must be a combination of a   capacitor   and   a
resistor because the time constant of the   aperiodic   component is relatively high.

Because there is only one passive component   in   every   section of the wiring the wiring diagram may be   this   (without
any information about the position   of   the   power   source).
                                                   (2 points)

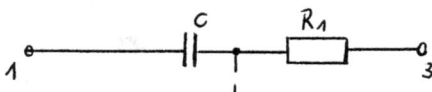

```
  o───────┤├───┬───[    ]───o
 1        C    |    R₁       3
              I
```

Measuring the a.c. you can also see that   there   must   be   a
capacitor between 1-3.

Repeating the measurments between 1-2 you will find the same results (qualitatively).

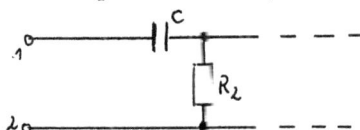

Using the same a.c. voltage $I_{13} > I_{12}$. Together with the result from 3. it follows that $R = R_1 + R_2$.   (2 points)

6. Position of the power source U_c.

Because there is no current between 1-2 without an external power source the wiring must be the following:

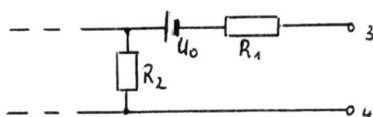

To proof the charging of the capacitor you have to discharge before!   (1 point)

The resulting wiring diagram is the following:

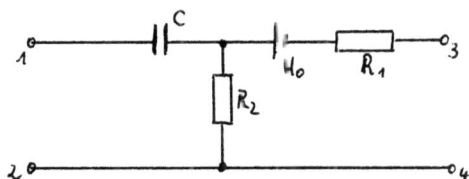

     (1 point)

7. Parameter determination of the components:

   R_1, R_2 and C (5 points)

   Results: $R = 1490 \,\Omega$, $R_2 = 463 \,\Omega$, $R = 1028 \,\Omega$,

   $Z_{12} = 1520 \,\Omega$, $Z_{13} = 1775 \,\Omega$, $C = 2.2 \,\mu F$.

   Z , Z : a.c. resistance between 1-2, 1-3.

8. Error estimation: (3 points)

The black box contens was:

U : battery   1.5 V,

R_1 : resistor   1 kΩ + 5%,

R_2 : resistor   470 Ω + 5%,

C : capacitor 2.2 μF + 5%.

     Total: (20 points)

HUNGARY

PHYSICS COMPETITIONS IN HUNGARY

Erzsébet Lugosi

(Mathematical and Physical Journal for High School Students)

and

Jenő Szép

(Loránd Eötvös University)

Description of Physics Competitions in Hungary

There are many different physics competitions in Hungary. In every year thousands of students take part in physics competitions in the whole country.

National competitions are organized for 15-18 year old students by the Ministry of Education. In these competitions there are three stages. In the first and second stage the students solve theoretical problems, in the third stage experimental ones.

In the Eötvös competition high school students or those who graduated from high school in the year of the competition may take part. In this competition there is only one stage with theoretical problems. The Eötvös competition was the first national physics competition in Hungary. First time it was organized in 1916 and was named after Irén Károly.

An important part of the physics education is the Mathematical and Physical Journal for High School Students - its popular name is KöMAL. It was founded by Dániel Arany in 1894. Nowadays this journal appears in more than 10,000 issues per month. In the KöMAL there are theoretical competitions and an experimental competition also. In every month the students solve three theoretical problems and/or an experimental one. They send the solutions in, and those are graded. At the end of the school year the best competitors are listed in the journal. The importance of this type of competition is that those students, who can work very well and successfully, but who does not have the ability of quick problem solving, may achieve good results. Many from the famous hungarian physicists and mathematicians state, that one of the reasons why they become scientists was the competition of KöMAL. All the problems and solutions below had appeared in the KöMAL in the 1982-1989 years.

A more than half century long Hungarian heritage of physics competitions and education through problem solving is the reason for the outstanding abilities of the hungarian students in problem solving. On the International Physics Olympiads the hungarian students received 14 first, 17 second and 36 third places so far. These achievements compared to the small population (10 million people) of Hungary are excellent.

Training of the Hungarian Team before the
International Physics Olympiad

In seven cities of Hungary special courses are hold
for those students who decided to work hard in order to
participate on the International Physics Olympiad. The
courses are on every second week during the school year. In
these courses the students learn about those areas of
physics which are outside of the subject of high school.
They solve difficult, complicated and interesting problems,
and discuss the useful methods for solving such problems.

At the end of the school year a 2-day competition is
organized for the best students of these courses, and for
those, who achieved good results at different physics
competitions. At the end of this competition it is decided
who from the students will participate on the IPhO.

Two weeks before the olympiad the selected students
meet for five days to prepare for the international
competition.

Few Problems for the Hungarian National Physics

Competition for 15 - 16 Year Old Students

1. A family wants to decide by vote whether they should make an excursion or not. They go if both children and one of the parents vote for it (the other parent has to stay at home to receive the post). Make plans for a voting machine consisting of four switches, a source and an electric bulb which lights up if they may go on the excursion. (The elements of the voting machine are shown in Figure 1.)

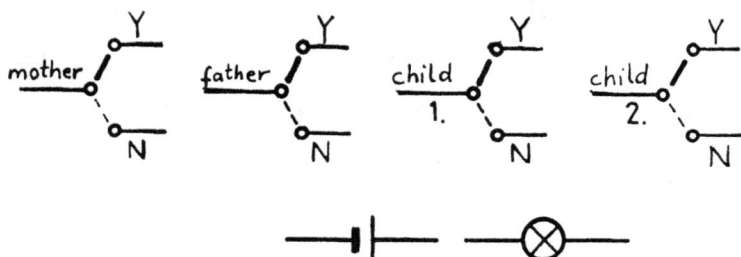

Figure 1.

Solution. The current may flow through that part of the circuit where the switches of the parents are only in the case if the parents vote differently. This happens if their switches are connected as in Figure 2. When the lamp

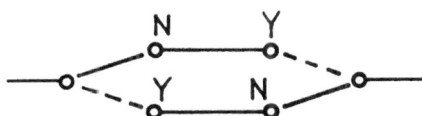

Figure 2.

lights, current flows through each children's part and
through the part of the parents. This means, the switches
of the children should be connected serially with the
system of the parents. The obtained voting machine can be
seen in Figure 3.

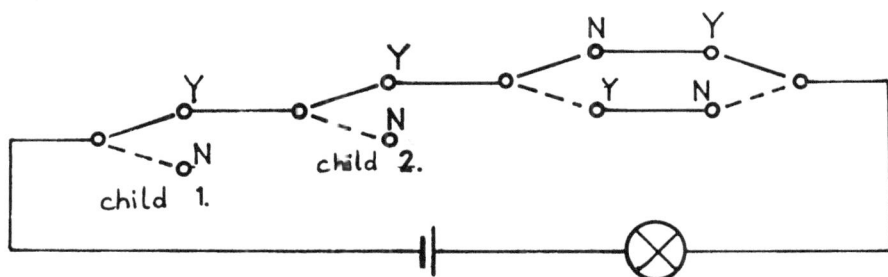

Figure 3.

2. The atoms in the crystal lattice of graphite are
arranged in parallel planes forming hexagonal grids in each
plane ("honey comb structure", see Figure 4). The distance
of the atoms in the planes is 142 pm. Find the distance
between the planes. (The density of graphite is
2270 kg/m^3.)

306

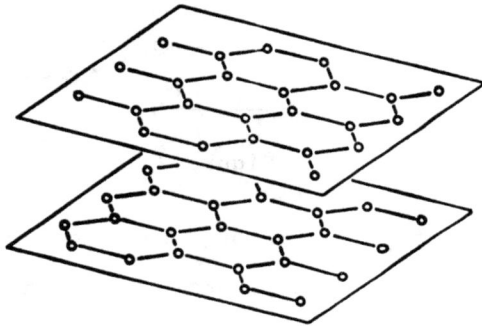

Figure 4.

Solution. Suppose that the distance of the planes is larger than the diameter of a carbon atom. In this case the planes in the crystal lattice can be displaced that way, that the atoms will be exactly above each other. The system of the atoms will look as in Figure 5.

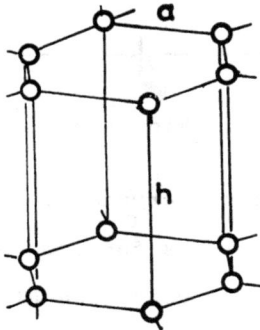

Figure 5.

Each atom belongs to six different cells, hence there are $12 \cdot (1/6) = 2$ atoms in one cell. The number of atoms in 1 m^3 graphite is

$$N = 6.023 \cdot 10^{23} \ (\text{atoms/mol}) \cdot 2.27 \cdot 10^6 \ g/(12 \ g/mol) =$$

$$= 1.14 \cdot 10^{29} \ \text{atoms}. \tag{1}$$

The number of cells is half of this value, that is $5.7 \cdot 10^{28}$. The volume of a cell is:

$$V = 1 \ m^3 \ / \ (5.7 \cdot 10^{28}) = 1.756 \cdot 10^{-29} \ m^3. \tag{2}$$

We should determine the height of a cell. The area of the base of a cell is

$$T = 3a^2 \ \sqrt{3} \ /2 = 5.239 \cdot 10^{-20} \ m^2, \tag{3}$$

where $a = 142$ pm. So the height of a cell, that is the distance between the planes is

$$h = V/T = 3.35 \cdot 10^{-10} \ m. \tag{4}$$

This value is larger than the diameter of a carbon atom, so the assumption we started with was adequate.

3. Determine the order of luminous intensity of the lamps shown in Figure 6.

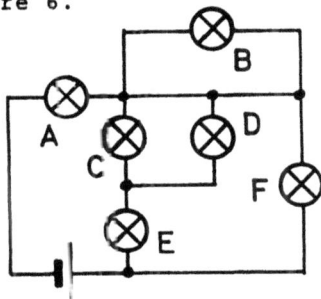

Figure 6.

Solution. Suppose that each lamp has the same resistance, which is constant, R. The power of a lamp is

$$P = I^2 R, \tag{5}$$

where I is the current flowing through the lamp. $I_B = 0$ because lamp B is in a short circuit. Neglecting this lamp the circuit is redrawn in Figure 7.

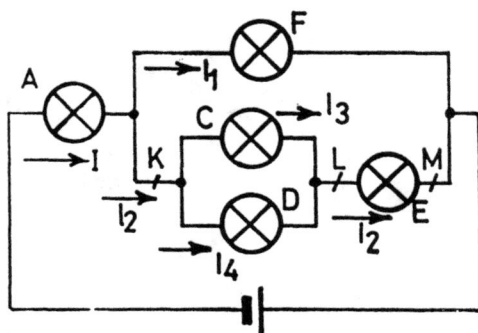

Figure 7.

The resistances between points K, L and K, M are:

$$R_{KL} = 1/(1/R+1/R) = R/2, \tag{6}$$

$$R_{KM} = R_{KL} + R_{LM} = (3/2)R. \tag{7}$$

Marking the currents as in Figure 7,

$$I = I_1 + I_2. \tag{8}$$

Because of the parallel connection

$$I_1/I_2 = (3/2)R/R = 3/2. \tag{9}$$

For I_1 and I_2 we obtain:

$$I_1 = (3/5)I, \qquad (10)$$

$$I_2 = (2/5)I. \qquad (11)$$

The current of lamp C and D is the same:

$$I_3 = I_4 = I_2/2 = (1/5)I. \qquad (12)$$

The current of E is I_2.

If P marks the power of lamp A then

$$P_A = P, \quad P_F = (9/25)P, \quad P_C = P_D = (1/25)P,$$
$$P_E = (4/25)P, \quad P_B = 0. \qquad (13-17)$$

The luminous intensity is larger if the power of the lamp is larger, hence the order of luminous intensity is the following: A; F; E; C = D; B.

4. Connecting a voltage of 0.4 V to the ends of a mercury column, contained by a thin glass tube, the current is 5 A. Calculate the current if we pour the mercury into another tube, which has one third the diameter of the former one, and connect the same 0.4 V on the ends.

(László Holics)

Solution. The resistance of the mercury in the original glass tube is

$$R_1 = \rho h/A, \tag{18}$$

where ρ is the specific resistance, h is the length and A is the cross-section of the mercury column. The resistance in the thinner tube, which cross-section is one-ninth of the previous tube:

$$R_2 = \rho \, 9h/(A/9) = 81 \rho h/A = 81 \, R_1. \tag{19}$$

The current in the thinner tube is:

$$I_2 = U/R_2 = U/(81R_1) = I_1/81. \tag{20}$$

That is the current is $5A/81 = 0.06$ A.

5. A dry uniform drinking straw is placed at right angles on the edge of a frictionless table in such a way that half of its length is protruding over the edge. A fly alights on the inner end of the straw and moves to the other end. The straw does not tip over. It does not tip over even when another fly settles down on top of the first. At most what can be the mass of the second fly?

(Imre Légrádi)

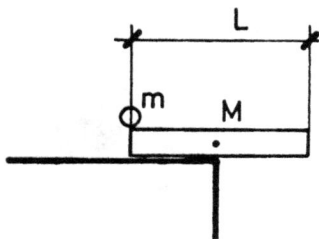

Figure 8.

Solution. In Figure 8 M, L, m marks the mass of the straw, its length, and the mass of the fly, respectively. The system of the straw and the fly is closed because the table is frictionless. The center of mass of this system will not change when the fly moves to the right. When the fly moves with a distance A to the right, the straw moves with a distance B to the left. Hence

$$MB = mA. \tag{21}$$

The fly moved to the end of the straw, hence

$$A + B = L. \tag{22}$$

From equations (21) and (22) we obtain:

$$B = Lm/(M+m). \tag{23}$$

a) If $M \leq m$ then $B \geq L/2$, that means the whole straw lies on the table, the mass of the second fly can be arbitrary.

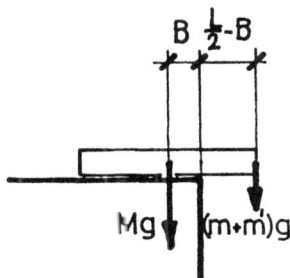

Figure 9.

b) In the case of m < M mark the mass of the second fly by m'. The condition of the equilibrium (see Figure 9) is

$$MqB \geq (m+m')g(L/2-B). \qquad (24)$$

Using equation (23)

$$m' \leq m(M+m)/(M-m). \qquad (25)$$

If M >> m then m' $\leq$ m, that is the mass of the second fly should be smaller than that of the first one.

6. On the electric meter one reads the following data:
220 V,   max 10 A,   2400 turns = 1 kWh.
Assume that the main voltage is constantly 220V. Give a prescription of how to use the meter for resistance measurement. What is the measuring range available?

(András Csordás)

Solution. The electric meter measures the energy taken from the current. The number of revolutions (n) of its disk is proportional to this energy:

$$W = \alpha \, n, \qquad (26)$$

where α is a coefficient and can be determined from the data:

$$\alpha=1 \text{ kWh}/2400 \text{ rev}=3.6 \cdot 10^6 \text{ J}/2400 \text{ rev} = 1.5 \cdot 10^3 \text{ J/rev}. \quad (27)$$

The voltage is constant, hence the resistance R through time t takes

$$W = P t = (U^2/R)t \qquad (28)$$

energy. The value of the resistance can be determined from (26) and (28):

$$R = U^2 t/(\alpha n). \qquad (29)$$

U and α are known. We need to count the number of rotations during time t, to calculate R.

The data "max 10A" means that at most a current of 10A can go through the electric meter, otherwise the fuse blows out. Because of this

$$I = U/R \leq 10A \qquad (30)$$

must be valid, therefore $R \geq 22\Omega$.

This is the minimal value of the resistance. When the resistance is larger than a certain value, the current is too small and is not enough to have the disk of the electric meter rotated. This upper limit is different for different electric meters.

7. A glass tube of small inner cross-section, length 1m, open at both ends, is half submerged in water. Covering the upper end we lift the tube and take it out of the

314

water. Find the length of the water remaining in the tube.

Solution. After submerging in water the tube contains water of 0.5 m length. Taking out of the water, a small portion of the water flows out, and the length of the remaining water is h. The pressure of the air inside the tube can be calculated by the Boyle - Mariotte law:

$$p = p_0 V/V_1 = p_0(L/2)A/[(L-h)A] = p_0 L/[2(L-h)],\qquad(31)$$

where L = 1 m and A is the cross section of the tube.

The pressure outside should be equal to the sum of the pressure of the water column and that of the air inside:

$$p_0 = p_0 L/[2(L-h)] + h\rho g,\qquad(32)$$

($\rho = 10^3$ kg/m^3 is the density of water). Solving this equation the physically acceptable value is h = 0.475m = 47.5 cm.

8. Given a uniform rod of mass m, length L, in vertical position, pivoted about a horizontal axis. Two horizontal springs of same force constant are fastened to the upper end of the rod as shown in Figure 10. Determine D, the force constant of the springs such that the rod should be in stable equilibrium.

(Jenö Szép)

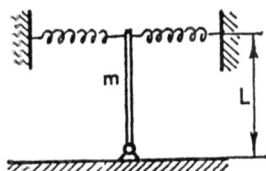

Figure 10.

Solution. A body is in stable equilibrium, if giving it a little push it returns to its original position. We should determine the conditions, under which the rod moves back to the vertical position. Mark by h the length of the springs, and by d their extensions. (If the springs are compressed, d is negative.) Remove the rod from its vertical position as seen in Figure 11 by an angle $\Delta\alpha$. If

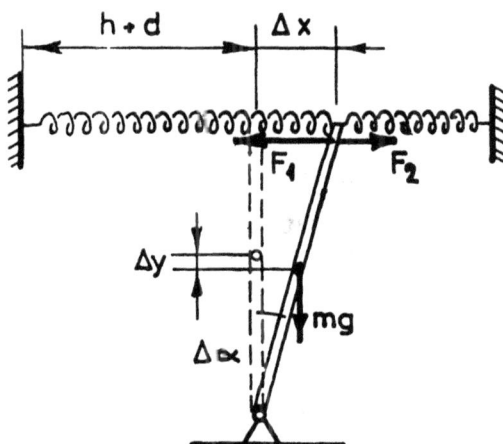

Figure 11.

$\Delta \alpha$ is very small, the position of the springs can be considered as horizontal. The extensions of the springs change by Δx, so the forces acting on the rod are:

$$F_1 = D(d + \Delta x); \tag{33}$$
$$F_2 = D(d - \Delta x). \tag{34}$$

The other forces which act on the rod are the force of gravity and the force of the flexible joint. The rod takes its original position if the torque of the forces makes this possible. So the condition of the stable equilibrium is:

$$D(d + \Delta x)L - D(d - \Delta x)L - mg\Delta x/2 > 0. \tag{35}$$

From this equation we obtain, that

$$D > mg/4L. \tag{36}$$

The rod will be in stable equilibrium if the force constant is larger than mg/4L.

Problems for the Hungarian National Physics

Competition in 1989 for 17 - 18 Year Old Students

Problems in the first stage

I/1. Two bodies of mass m_1 and m_2 start to fall from height h just after each other (see Figure 12). All collisions happen to be on a vertical axis and are absolutely elastic.

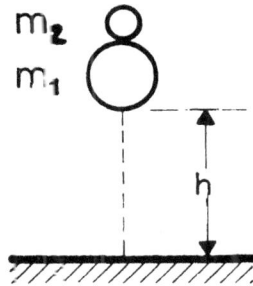

Figure 12.

a) For what ratio of the masses will the body of mass m_1 stay in equilibrium after the collisions?

b) Determine the raising's height of the body of mass m_2 in this case.

(Péter Ungár)

Solution. Each body reaches the ground with velocity $v_0 = -\sqrt{2gh}$. The body of mass m_1 starts to move upwards with velocity v_0 and collides with the body of mass m_1 falling down with velocity $-v_0$. The velocities after the collision are

$$u_1 = v_0(m_1-3m_2)/(m_1+m_2), \tag{37}$$

$$u_2 = v_0(3m_1-m_2)/(m_1+m_2). \tag{38}$$

a) The body of mass m_1 stays in equilibrium after the collision, that is $u_1 = 0$ if $m_1/m_2 = 3$.

b) In this case $u_2 = 2v_0$, the raising's height of the body of mass m_2 is 4h.

I/2. In Figure 13 the distance between points A and B is 14 m, the length of each pieces of rope is 5 m. Two bodies of mass 7 kg are hanging in points C and E. In point D a body of appropriate mass m is hanging in such a way, that the point D is in 7 m below the line AB. The body of mass m is slowly raised until points D, C and E will be on a line. Determine the work done.

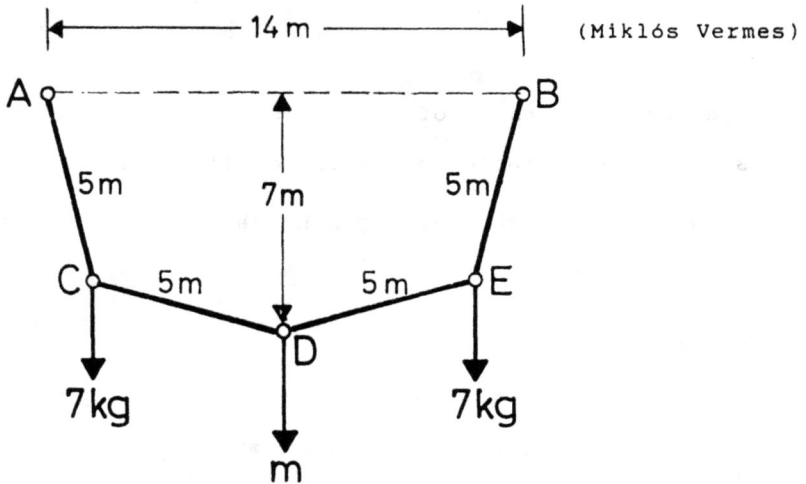

(Miklós Vermes)

Figure 13.

Solution. **First we** determine the value of mass m. (See Figure 14.)

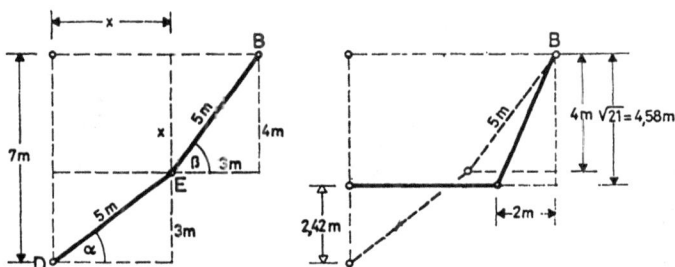

Figure 14.

From the data

$$(5m)^2 = x^2 + (7m-x)^2, \qquad (39)$$

hence x=4m, tgα = 3/4, tgβ = 4/3. In the first part of the rope (DE) the force is mg/(2sinα) in the second part (EB) is (m/2+7kg)g/sinβ. The values of the horizontal components of these forces are equal in point E:

$$((m/2+7kg)g/sin\beta)cos\beta = (mg/(2sin\alpha))cos\alpha. \qquad (40)$$

Solving this equation we obtain m = 18kg.

From Figure 14 it can be seen, that the body of mass 18kg is raised by 2.42m, the increase in potential energy at this body is (18kg)(2.42m)g = 427 joule. The position of the bodies of mass 7kg changed by 0.58m, hence the decrease in potential energy at these bodies is 2·(7 kg)·(0.58 m)·g = 80 joule.

The work done is 427 joule - 80 joule = 347 joule.

I/3. Two condensators of 2 µF and 3 µF are charged
to 150 V and 120 V, respectively. Two plates of the
condensators are connected as in Figure 15. Two wires are
going out freely from the other plates of the condensators.
A discharged condensator of capacity 1.5 µF falls to the
free endings of the wires.

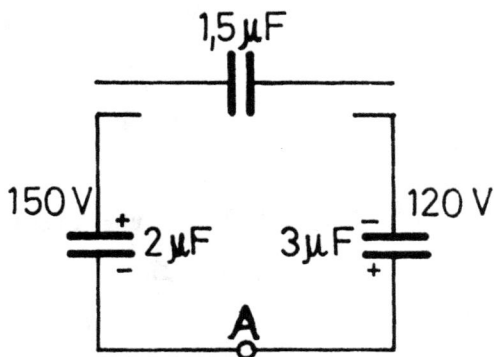

Figure 15.

a) Calculate the charge on the condensators after
this.

b) How much charge flows through point A, and in which
direction?

(Miklós Vermes)

Solution. There will be x and -x charges on the plates

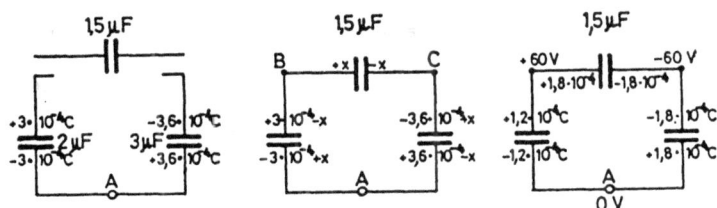

Figure 16.

of the condensator of capacity 1.5 μF. In Figure 16 the
left hand side figure shows the original state; in the
middle figure the new charges can be seen. The voltage
between the points B and A calculated through the
condensator of capacity 2μF and through the other two
condensators give the same values:

$$(3 \cdot 10^{-4} C - x)/(2 \cdot 10^{-6} F) =$$
$$= -(3.6 \cdot 10^{-4} C - x)/(3 \cdot 10^{-6} F) + x/(1.5 \cdot 10^{-6} F). \qquad (41)$$

The solution of this equation is $x = 1.8 \cdot 10^{-4}$ C.

a) The potential (compared to the point A) on the
condensator of capacity 2uF is +60V, on the condensator
of capacity 3μF is -60V, the potential difference between
the two plates of the third condensator is 120 V.

b) $1.8 \cdot 10^{-4}$C charge flowed through point A from the
right to the left.

The right hand side figure in Figure 16 shows the
charges of the condensators and the potentials.

I/4. A glass tube of length 76 cm and closed upper end dips into mercury. The mercury fills a portion of the tube. There is 0.001 mol air in the closed volume (see Figure 17).

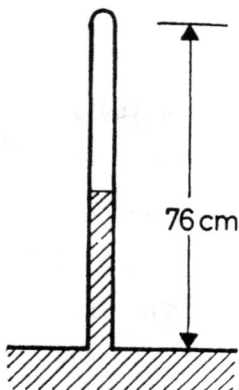

Figure 17.

The air-pressure outside keeps equilibrium with a mercury-column of height 76 cm. The molar heat of the air at constant volume is $C_v = 20.5$ J/(mol K). Calculate the heat-loss of the air closed in the tube when its temperature decreases by 10 ^{o}C.

(Ervin Szegedi)

Solution. On the basis of the first law the heat gained by the gas is

$$\Delta Q = \Delta E - \Delta W. \qquad (42)$$

The changing of internal energy:

$$\Delta E = \mu c_v (T_2 - T_1) = 0.001 \cdot 20.5 \cdot (-10) = -0.205 \text{ joule.} \quad (43)$$

There are different ways to calculate the work done:

a/ The pressure of the closed air equals to the pressure of the mercury of the same height as the air column, because this is the pressure which caused the mercury column of height 76 cm to sink. The work done is $\Delta W = p \Delta V$. In our case the pressure changes linearly with the volume, hence the pressure can be taken as the average of the pressure at the beginning (p_1) and at the end (p_2)

$$p = (p_1 + p_2)/2. \quad (44)$$

The change in volume is

$$\Delta V = V_2 - V_1. \quad (45)$$

The work done on the gas:

$$\Delta W = -(1/2)(p_1 + p_2)(V_2 - V_1), \quad (46)$$

$$\Delta W = -(1/2)(p_2 V_2 - p_1 V_1 + p_1 V_2 - p_2 V_1). \quad (47)$$

The algebraic sum of the third and fourth terms equals to zero, because (see Figure 18)

$$p_1 V_2 = \rho g x A y \quad \text{and} \quad p_2 V_1 = \rho g y A x, \quad (48-49)$$

(the base of the tube is A, the density of mercury is ρ) and these values are equal to each other.

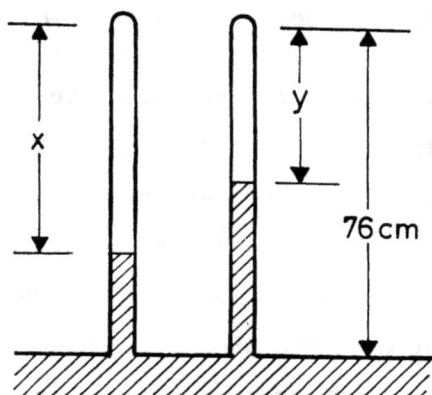

Figure 18.

Using the gas law for the first and second terms:

$$p_2 V_2 = \mu RT_2; \quad p_1 V_1 = \mu RT_1. \qquad (50-51)$$

The work done on the gas:

$$\Delta W = -(1/2)\mu R(T_2 - T_1) = 0.042 \text{ joule.} \qquad (52)$$

b/ During the process the pressure and the volume are proportional (Figure 19). The pressure is ρgx, the volume is Ax, according to the gas law:

$$pV = \rho gxAx = \rho gAx^2 = \mu RT. \qquad (53)$$

The work done can be calculated from the area of the trapezoid in Figure 19.

$$\Delta W = -(1/2)(p_1 + p_2)(V_2 - V_1) = -(\rho gA/2)(x_2^2 - x_1^2). \qquad (54)$$

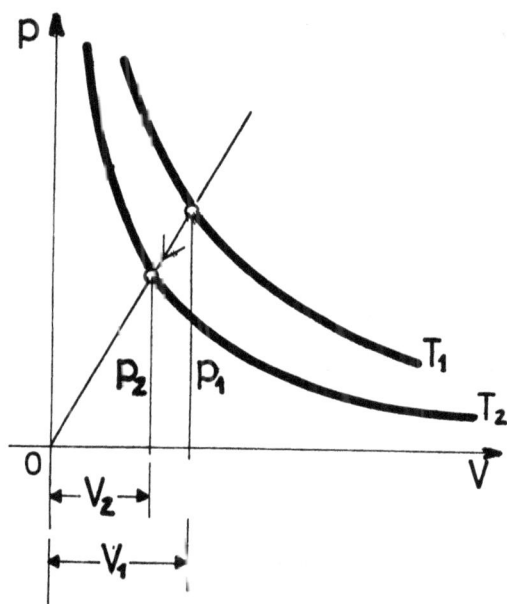

Figure 19.

At the cross points:

$$\varrho g A x_1^2 = \mu R T_1, \qquad x_1 = \sqrt{\mu R T_1 / \varrho g A}, \tag{55}$$

and

$$\varrho g A x_2^2 = \mu R T_2, \qquad x_2 = \sqrt{\mu R T_2 / \varrho g A}, \tag{56}$$

The work done on the gas is:

$$\Delta W = (\varrho g A/2)(\mu R T_2/\varrho g A - \mu R T_1/\varrho g A) = -(\mu R/2)(T_2 - T_1). \tag{57}$$

Knowing the work done the heat increase of the air can be calculated:

$$\Delta Q = \mu C_v(T_2 - T_1) + \mu(R/2)(T_2 - T_1)$$

$$= -0.205 \text{ joule} - 0.042 \text{ joule} = -0.247 \text{ joule}. \quad (58)$$

It is interesting that the result is independent from the initial value of x.

II/1. A thin ring of radius r = 10 cm, rotating in a horizontal plane falls to a table from a height h = 20 cm (see Figure 20).

Figure 20.

The ring rotates around its vertical axis in the air with an angular velocity ω_0 = 21 s^{-1}. The collision is nonelastic and takes very short time. The coefficient of friction between the table and the ring is μ = 0.3. Calculate the number of turns done by the ring while it stops.

(Ervin Szegedi)

Solution. The falling takes

$$t_0 = \sqrt{2h/g} = 0.2 \text{ sec} \quad (59)$$

time, the velocity at the table is

$$v = gt_0 = 2 \text{ m/s}. \tag{60}$$

The ring turns by angle

$$\varphi_0 = \omega_0 t_0 = 4.2 \tag{61}$$

while falling.

During the very short time of the collision the force between the ring and the table is $mv/\Delta t$, the frictional force is $\mu mv/\Delta t$, its torque is $\mu mvr/\Delta t$. The time of collision is very short, hence the weight of the ring can be neglected at the calculation of the frictional force. The moment of inertia of the ring is $\theta = mr^2$. The angular acceleration is

$$\beta = \mu rmv/\theta \Delta t = \mu v/r \Delta t. \tag{62}$$

The decrease in the angular velocity during the collision within time Δt is

$$\Delta\omega = \beta \Delta t = \mu v/r = 6 \text{ s}^{-1}. \tag{63}$$

Because of this small time period the turn of the ring can be neglected.

Right after the collision the angular velocity is

$$\omega_1 = \omega_0 - \mu v/r = 15 \text{ s}^{-1}. \tag{64}$$

The rotation of the ring on the table will slower as the result of the force μmg. The torque of this force is μmgr.

The angular acceleration is

$$\beta_1 = \mu mgr/\theta = \mu g/r = 30 \ s^{-2}. \tag{65}$$

The ring will stop within

$$t_1 = \omega_1/\beta_1 = 0.5 \ sec \tag{66}$$

time. The total angle of rotation during this time is $\beta_1 t_1^2/2 = 3.75$.

The total rotation is $4.2 + 3.75 = 7.95$, the number of turns done is $7.95 / 2\pi = 1.265$.

II/2. A cylinder with a piston contains 4 g of helium and 16 g of oxygen of temperature 0^oC and pressure 10^5 Pa. (See Figure 21.)

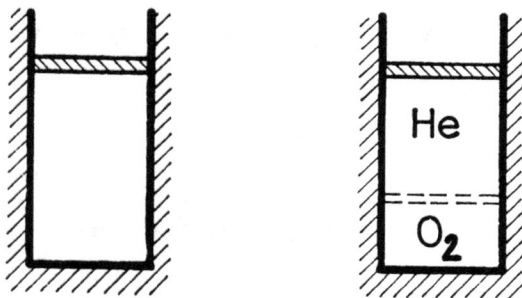

Figure 21.

The walls of the cylinder and the piston are heat-insulators. The pressure is increased to $2 \cdot 10^5$ Pa. Determine the temperature and the volume of the gas-mixture

after this. The molar heats for helium are: C_{vh} =
= 12.3 J/molK, C_{ph} = 20.5 J/molK;   for   oxygen: C_{vo} =
= 20.5 J/molK, C_{po} = 28.7 J/molK.

(Miklós Vermes)

Solution. The helium and the oxygen has different
specific heat constant (κ), hence the adiabatic law
(pv^{κ} = const) determines different volumes (and
consequently different temperatures) of these gases when
their pressures change on the same way.

Suppose in mind that we place a heat insulator piston
between the separated oxygen and helium. First we determine
the volumes and temperatures of the separated gases.

In the case of helium V_{h1} = 22.4 dm^3, κ_h = 5/3, its
new volume is V_{h2}. The adiabatic law for this gas is
$10^5 \cdot 22.4^{5/3}$ = $2 \cdot 10^5 \cdot V_{h2}^{5/3}$. From this equation the volume
of the compressed helium is V_{h2} = 14.778 dm^3. The
temperature of the helium can be calculated from the gas-
law:

$$10^5 \cdot 22.4/273 = 2 \cdot 10^5 \cdot 14.778/T_{h2}, \qquad (67)$$

Hence T_{h2} = 360.26 K.

In the case of oxygen V_{o1} = 11.2 dm^3, κ_o = 1.4, its
new volume V_{o2}. The adiabatic law for this gas is
$10^5 \cdot 11.2^{1.4}$ = $2 \cdot 10^5 \cdot V_{o2}^{1.4}$. From this equation the volume
of the compressed oxygen is V_{o2} = 6.826 dm^3. The
temperature can be calculated from the gas-law:

$$10^5 \cdot 11.2/273 = 2 \cdot 10^5 \cdot 6.826/T_{o2}, \tag{68}$$

hence $T_{o2} = 332.77$ K.

Now we remove in mind the heat insulator piston. The two gases will mix at a constant pressure and the common temperature will be T.

$$1 \cdot 20.5 \cdot (360.26-T) = 0.5 \cdot 28.7 \cdot (T-332.77). \tag{69}$$

From this equation the temperature is T = 348.94.

The volume of helium in the mixture is:

$$10^5 \cdot 348.94 \cdot 22.4 \ / \ (2 \cdot 10^5 \cdot 273) = 14.314 \ dm^3. \tag{70}$$

The volume of oxygen in the mixture:

$$10^5 \cdot 348.94 \cdot 11.2 \ / \ (2 \cdot 10^5 \cdot 273) = 7.157 \ dm^3. \tag{71}$$

The total volume of the mixture is 14.314 + 7.157 = = 21.471 dm^3.

While mixing with the oxygen the helium gave kinetic energy to the oxygen, because the molecules of the oxygen need more energy on the same temperature than the molecules of the helium, because they rotate also.

Note. The Poisson-law of the adiabatic change of a gas-mixture is the following: If we have n_1 mol gas with coefficients C_{p1}, C_{v1} and n_2 mol gas with coefficients C_{p2}, C_{v2} then

$$p V^{(n_1 C_{p1}+n_2 C_{p2})/(n_1 C_{v1}+n_2 C_{v2})} = const. \tag{72}$$

II/3. A closed "black-box" is given. Turning alternate voltage of different frequency to the two outcoming wires the following AC-impedances are obtained:

ω [s^{-1}]	20	200	250	300	325	350	400	1000	5000
Z [Ω]	782	53.0	34.0	25.4	25.2	27.2	34.9	145.5	792

What is inside the box?

(Miklós Vermes)

Solution. The AC-impedance has a minimum, therefore the box contains a condensator and a coil which are in a

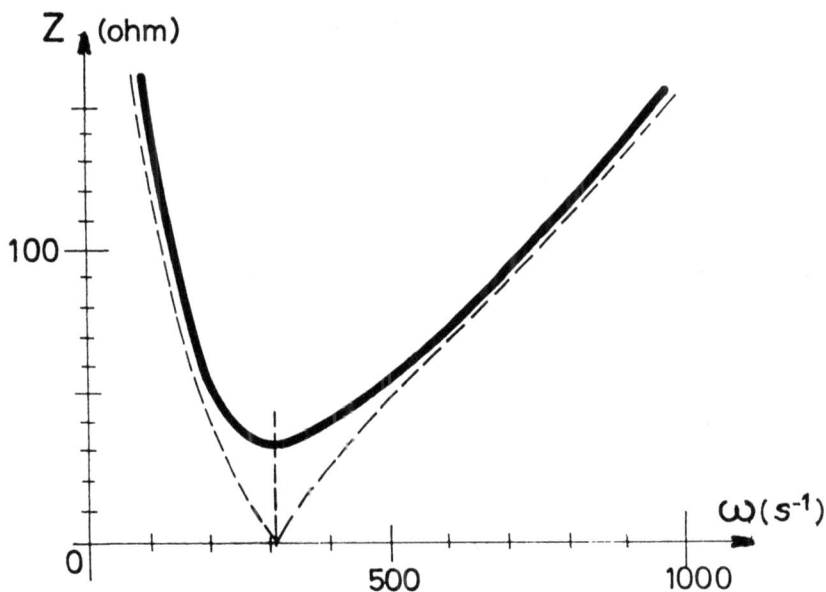

Figure 22.

serial connection (Figure 22). The AC-impedance is not zero at the resonance frequency, therefore one ohmic resistance should be connected in series also. With inductivity L, capacity C, and resistance R connected in series the resistance of alternate current will be:

$$Z = \sqrt{R^2 + \frac{(1- \omega^2 LC)^2}{(\omega C)^2}} \; . \tag{73}$$

Using three data pairs from the table above we could obtain three equations, and solve them. The calculations would be pretty tiering, we can obtain the L, C, R values with an adequate accuracy on an easier way.

At resonance Z = R. The minimal value of the impedance is R = 25 ohm. So this is the value of the ohmical resistance. (The frequency at the resonance is ω_r = = 310^{-1}.) At a very small frequency essentially the condensator is the only source of indepence:

$$782 = 1/ \; 20C, \tag{74}$$

hence C = 63.9μF.

At very large frequency essentially the coil only causes impedance:

$$792 = 5000L, \tag{75}$$

hence L = 0.158 henry.

Experimental Problems for the Competition of

KöMAL for 15 - 18 Year Old Students

1. In fluids and gases the motion of a body is impeded by drag force. This force is made up of two components. One of them is the result of internal friction and it is proportional to the velocity of the body. The other force is caused by the vortices behind the body and it is proportional to the square of the velocity. Determine by measurement which force is more important in the case of a balloon falling free in air.

Solution. If the balloon falls from very high we can observe, that first it moves accelerating, then uniformly. When the balloon falls uniformly the sum of the forces acting on it should be zero. That means, the force of gravity is the opposite of the drag force. The constant velocity of the balloon can be determined by measuring the distance it covered and the time. The connection between the velocity and the drag force can be determined by changing the weight of the falling object. We can do that by suspending small weights on the balloon.

The average velocity and drag force values obtained by measurements are plotted in Figure 23. To answer the question of the problem, we should determine, whether these

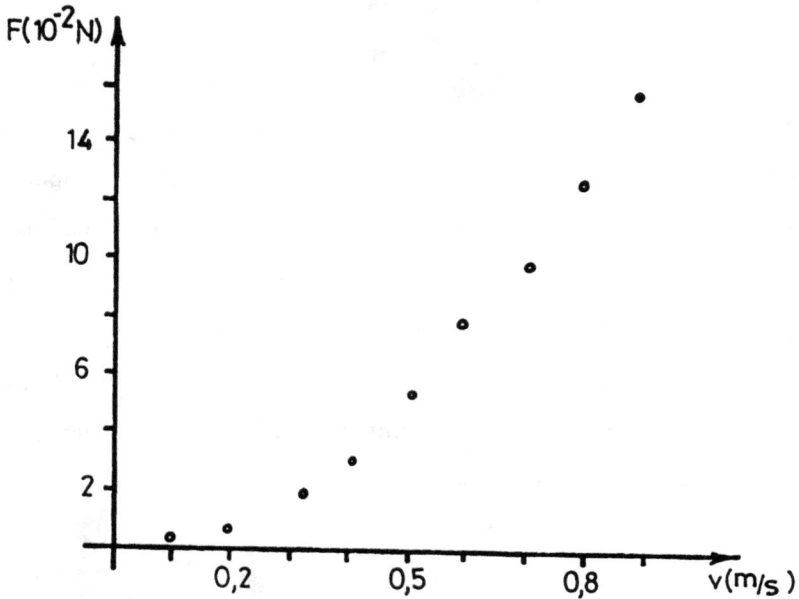

$F(10^{-2}N)$

14

10

6

2

$0,2$ $0,5$ $0,8$ $v(m/s)$

Figure 23.

points are close to a first order or a second order function. To decide it we should see the expression for the force which is the result of internal friction

$$F_f = 6\pi \eta r v, \tag{76}$$

and for the force caused by the vortices behind the body:

$$F_v = c \rho r^2 v^2, \tag{77}$$

(η is the viscosity of the air, r is the radius of the balloon, v is its velocity, ρ is the density of the air, c is a constant depending on the shape of the body.)

Using the SI system in expressions (76) and (77) we

may take their logarithm:

$$\lg F_f = \lg rv + \lg 6\pi\eta, \tag{78}$$

$$\lg F_v = 2 \lg rv + \lg c\rho. \tag{79}$$

Plotting $\lg F_f$ and $\lg F_v$ as function of $\lg rv$, the lines what we obtain have a slope of 1 and 2, respectively.

If we plot now the logarithm of the measured drag force as function of $\lg rv$, only the slope of the line should be determined.

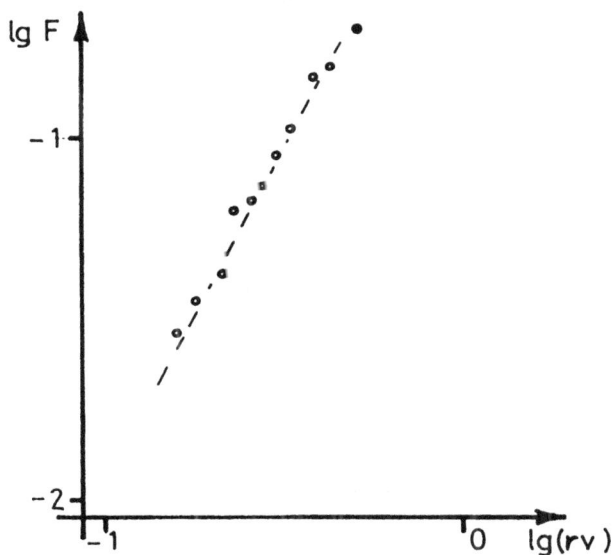

Figure 24.

The values plotted in Figure 24 were obtained by Krisztián Tar student. The points are close to a line which slope is 2. So we can say that the force caused by the vortices behind the body is more important in the case of a falling balloon.

2. Measure the downward bend of a stickpin as function of the bending force first for a new pin then after having tempered it in the flame of a match (raising the pin to heat and then cooling it.)

(Géza Tichy)

Solution. The tip of a stickpin was fixed by squeezing in a vise. A dynamometer was joined to the head of the pin and different forces were used through the dynamometer. To get a better measurement on the movement of the head of the tip it was projected to the wall, and the distance was measured on the shade.

The values what János Steiber student measured for 5 different pins are in Table 1 for new pins and for tempered ones. The table contains the average values and the variance also.

Table 1.

Force (N)	Bend of a stickpin (mm)													
	new pin							tempered pin					
	1	2	3	4	5	aver.	var	1	2	3	4	5	aver.	var
0.5	1	1	1	1	1	1.0	0.0	2	2	2	3	4	2.5	0.8
1.0	2	2	1.5	2	3	2.1	0.5	8	10	8	9	8	8.6	0.8
1.5	3	4	2	3	4	3.2	0.7	12	13	11	11	14	12.2	1.2
2.0	5	7	3	5	6	5.2	1.3	–	–	–	–	–	–	–
2.5	8	10	5	7	9	7.8	1.7	–	–	–	–	–	–	–
3.0	12	14	10	10	14	12.0	1.8	–	–	–	–	–	–	–

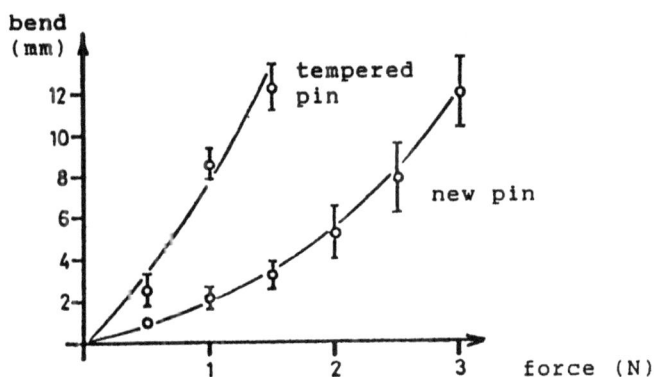

Figure 25.

The values of Table 1 are plotted in Figure 25. In this figure we can see, that the new pin bends less than the tempered one for the same force.

3. Investigate the tensile strength of drawing paper as function of moisture-content.

Solution. For the measurement paper sheets cut to the shape of Figure 26 were used.

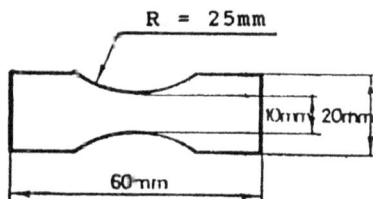

Figure 26.

Using this shape a paper sheet tears at a well determined place, where it is the narrowest. The equipment used can be seen in Figure 27.

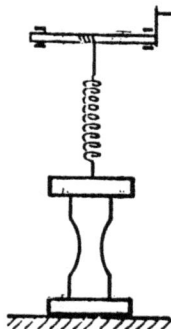

Figure 27.

The force was measured by the extension of the spring, which was calibrated before the measurement.

To obtain differently wetted papers a dry paper was placed between moistured blotters for different length of time. The moister content of a paper was measured in weight percentage.

The accuracy of measuring the force was 0.1 N. During the measurement the paper was slowly drying. The accuracy of measuring the moister content was 2%.

The measurement was done that way also, that a fully moistured paper was dried partially to obtain a dryer one.

In these two cases Itala Kucsera student measured the values of Table 2 and Table 3.

Table 2.

Table 2.
Measurements starting with a dry paper (A)

Moisture (%)	0.0	4.0	7.6	12.0	16.1	25.8	36.0	44.0	51.5
Force (N)	22.8	21.0	16.8	11.1	7.5	2.3	2.1	2.0	1.5

Table 3.
Measurements starting with fully moistured paper (B)

Moisture (%)	51.5	39.C	30.5	24.0	16.0	10.5	6.5	0.0
Force (N)	2.0	2.0	2.0	2.6	3.8	8.2	14.5	20.5

The values of Table 2 and Table 3 are plotted in Figure 28.

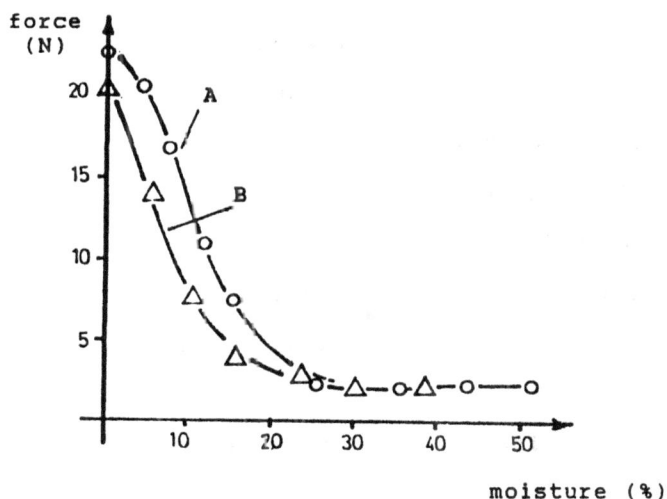

Figure 28.

From this figure we can see, that the tensile strength

of a drawing paper quickly decreases by wetting it. The reason is that the water mellows the fibers. The fully moistured and then dried paper is less strong than that one which was only moistured. That shows that the already weakened fibers cannot regain their original strength.

4. Drive a sufficiently big screw into soft wood. First you should drill a hole, its diameter should be about half of the diameter of the screw. Determine experimentally how the necessary torque changes while driving in the screw.

(Géza Tichy)

Solution. To determine the torque we must measure the value of the acting force on a given force-arm. This measurement can be done on many different ways.

a) If we drive the screw into a fixed wood, we can do the following: We fix a stick to the screw, and at the end of the stick drive the screw using a dynamometer. This system can be seen in Figure 29. If the force act in right angle on the stick, the torque is the product of the force shown on the dynamometer, and the length of the stick. Driving continuously the screw, it is difficult to keep the exact right angle at the end of the stick. We can change our system that way that this angle will be "automatically" a right angle.

Figure 29.

b) Fixing the dynamometer and driving the wood itself, the angle is fixed, and it is approximately a right angle (see Figure 30).

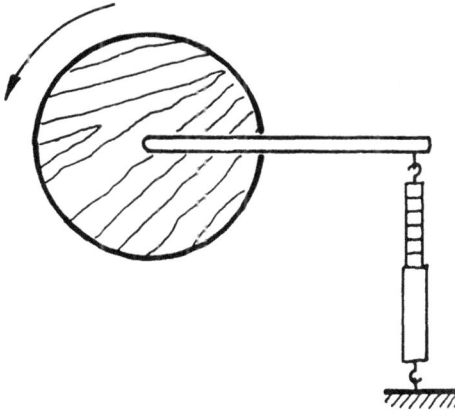

Figure 30.

c) Now we fix a disk to the screw and wind a thread to its edge. We pull the thread by the dynamometer. This arrangement can be seen in Figure 31.

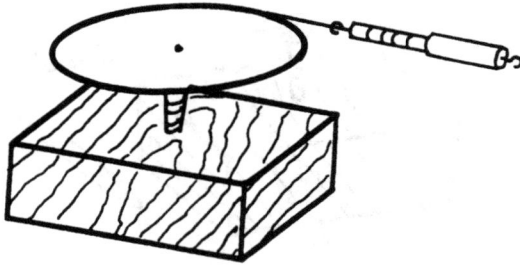

Figure 31.

Figure 32.

d) There are other similar equipments in Figure 32, and in Figure 33. In these cases the screw is fixed.

At this measurement one should be careful about the followings:

- Driving a screw more than once into the same hole, the torque will be smaller, because the hole is getting larger.

- On the other hand the used screw behaves differently

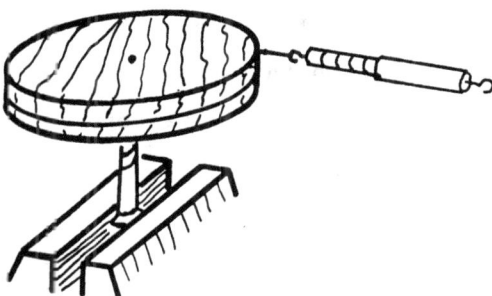

Figure 33.

than a new one.

- If something happens to the wood (e.g. it cracks)
that also will change the results of the measurement.

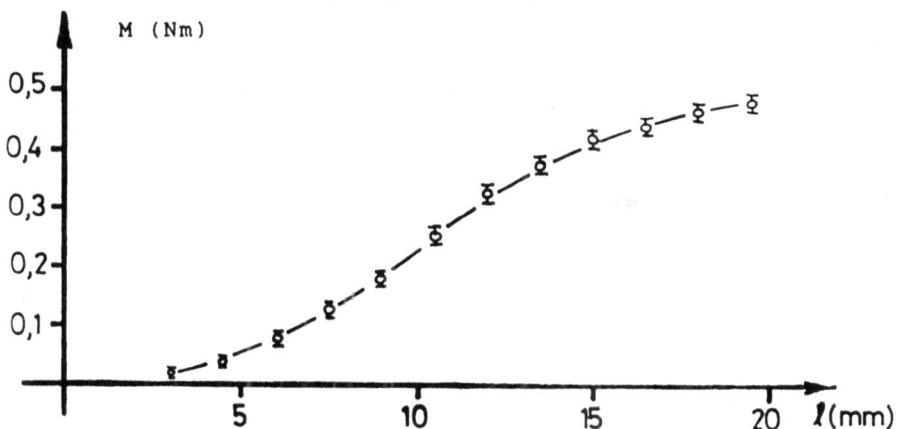

Figure 34.

Using different type of screws and woods we obtain
different results. Plotting the torque (M) as function of
the deepness of the screw in the wood we obtain similar
graphs for the different cases. Using a screw of length

23 mm, diameter 3 mm, and a pine wood Itala Kucsera student obtained the graph in Figure 34. The errors - which are the consequences of the facts mentioned before - are marked also in this figure.

5. Determine by measurement the pressure in a given balloon as a function of the radius.

(Erzsébet Lugosi)

Solution. When blowing up first a given balloon its material expands, and remains partially expanded after going down. This happens because of the changing of the elasticity of the rubber. Because of this the function we want to determine will be different for blowing up and going down. Also, if we blow up a balloon second time, the function we obtain will be different than it was at the first blowing.

In the following the size and pressure will be measured at the first blowing up and going down.

The device used in the experiment can be seen in Figure 35. There was water in the retort, and a thin glass tube was placed in to measure the pressure. (The pressure can be calculated from the height of the water.) The other ending of the retort was attached to the mouth of the balloon. The radius was measured on the shade of the balloon on the wall.

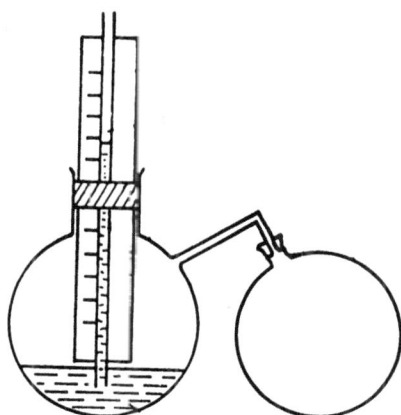

Figure 35.

The balloon burst above radius 13 cm, so the measurement was done in the range 0 - 13 cm. The average of the measured values for different balloons at blowing up and going down can be seen in Table 4 and Table 5.

Table 4.
Blowing up (A)

r(cm)	6	7	8	9	10	11	12	13
p - p_0 (Pa)	2190	2100	2050	2000	2030	2150	2200	2400

Table 5.
Going down (B)

r(cm)	13	12	11	10	9	8	7	6
p - p_0 (Pa)	1680	1300	980	800	700	640	460	410

These values were measured by Mátyás Szövényi-Lux
student. The graph can be seen in Figure 36. The error was
approximately 20%. It is large because of the different
material of the balloons used.

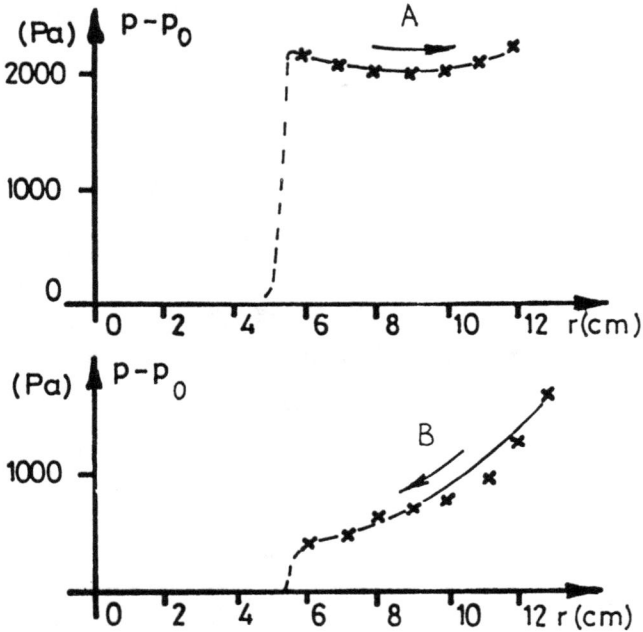

Figure 36.

The difference between blowing up and going down can
be seen very well in the figure. When the radius is smaller
than 6 cm the pressure cannot be measured because the
rubber is not tight. The broken lines show these parts of
the functions in the figure.

R. I. IRAN

DESCRIPTION OF THE IRANIAN COMMITTEE OF PHYSICS OLYMPIAD

1 Introduction

The Deputy Ministry of Education who is the head of the Organization for Research and Educational Planning is responsible for holding Physics Olympiad Compititions in Iran. This Deputy Ministry has set up a committee named Iranian Committee of Physics Olympiad, which is authorized in making all of the decisions regarding IPO affairs. This Committee is composed of university professors, high school physics teachers and executive directors. The following problems are discussed and decided about in this committee:

1. How to hold the preliminary competition among the third grade volunteer students. (Secondary Education in the Islamic Republic of Iran takes four years of studies).

2. Selecting the appropriate number from the volunteers.

3. Developing an appropriate curricula for the accepted students in the scope of fourth grade programs.

4. Holding the semifinal competition among the accepted students.

5. Organizing a few weeks complementary course for the selected students in the semi final competition.

6. Selecting the final members of the Team.

7. Carrying out all the financial and administrative tasks and dispatching the Team to IPO.

1.1 Programme of Training of the Team before the IPO

In the last days of each academic year(mid of June), a double stage preliminary competition is held among the third grade students of high schools. The first part of the competition consists of about 50 multiple choice questions. The second part is a written exam with about 8 problems to be solved.

Both of the exams are from the textbooks materials, which students have learned about in the high school. (It should be mentioned that curriculum development and textbook compiling is a centralized and uniform activity in the Islamic Republic of Iran).

The number of students participating in this preliminary competition is about 3000, from which the top 30s will be selected. Since only 5 students from these 30 will be dispatched to the IPO, and the rest should participate in the university entrance exam; the complete program of fourth grade of high school would be launched for them, along with some changes in physics curriculum. They start their fourth grade in September, the same as other regular students. After six months of study, a semi final theoretical and practical exam will be held, through which seven students will be selected and the rest will leave the course to participate in the high school final exams.

The seven selected students will enjoy from the following advantages:

a- They are permitted to take the final exams of high school in September, instead of June.

b- They are exempted from taking the university entrance exam, so as to be able to continue their higher education.

2 THEORETICAL PROBLEMS & THEIR SOLUTIONS

I-Some of the multiple choice questions given in the first stage of the preliminary competition in June 1989. Their answers are tabulated at the end of this part.

#1 If the elongation of each of the identical spring in fig(1) is 4 cm, what is their elongation if they are connected in series, with the same body suspended from the lower one

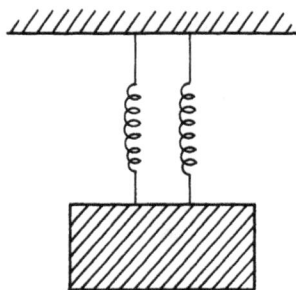

Fig(1)

a) 2 cm

b) 4 cm

c) 8 cm

d) 16 cm

#2 An iron block is floating on mercury. If we lower the temperature of the system from $30C°$ to $15C°$, the volume of the floating part will.

a) Increase

b) Decrease

c) Not change

d) The given data in not sufficient

#3 Consider a rectangular cube on a horizontal surface as shown in fig.(2). Find the min. height that we can raise MN edge so that the cube rotates around OP and lies on the ajacent face.

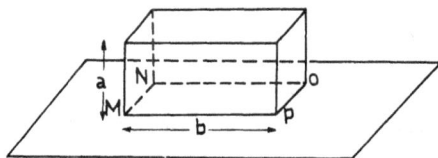

Fig(2)

a) $\frac{b^2}{\sqrt{a^2+b^2}}$

b) $\frac{a^2}{\sqrt{a^2+b^2}}$

c) b

d) $1/2\sqrt{a^2+b^2}$

#4 An empty plastic bag of weight P measured by a dynamometer, is filled with air at the atmospheric pressure and weighted again by same dynamometer. If the weight of air is P', the second measurement will show

a) $P - P'$ b) $P + P'$ c) P d) P'

#5 The length of a copper rod is to measured by an iron stick. When the temperature of surrounding is θ_1, the length of rod is L. What is its length, when the temperature of surrounding is $\theta_2(\Delta\theta = \theta_2 - \theta_1)$

a) $L\{1 + (\lambda_{Fe} - \lambda_{Cu})\Delta\theta\}$ b) $L\{1 + (\lambda_{Cu} - \lambda_{Fe})\Delta\theta\}$

c) $L\{1 + (\lambda_{Cu} + \lambda_{Fe})\Delta\theta\}$ d) $L\{1 + \lambda_{Cu}\Delta\theta\}$

#6 A thermometer whose calibration is not specified shows 5 ^{C}o as 50, and -20 ^{C}o. as 10. At what temperature it will read the same number as a thermometer with celsius calibration

a) - 40   b) +30   c) -70   d) None

#7 The internal energy of substance A and B are equal and substance A is in thermal equilibrium with substance C. Which of the statements is necessarily correct?

a) B and C are in thermal equilibrium.
b) The internal energy of B and C are equal.
c) If A and B are alike, B and C are in thermal equilibrium.
d) Temperatures of A and B are equal.

#8 The lengths of two metalic rods at temperature θ are L_A & L_B and their linear coefficients of expansion are λ_A & λ_B respectively. If the difference in their lengths is to remain constant in any temperature,

a) $\frac{L_A}{L_B} = \frac{\lambda_A}{\lambda_B}$

b) $\frac{L_A}{L_B} = \frac{\lambda_B}{\lambda_A}$

c) $\lambda_A = \lambda_B$

d) $\lambda_A \lambda_B = 1$

#9 In a closed glass tube filled with oil, there is an air bubble.
Which of the following statements is correct?

a) If we heat the tube,the volumes of oil and air bubble will increase.
b) If we cool the tube,the volumes of oil and air bubble will increase.
c) If we heat the tube the volume of oil will increase and volume of air
bubble will decrease.
d) If we cool the tube the volume of oil will increase and volume of air
bubble will decrease.

#10 A wooden stick stands vertically on the ground and sun rays fall on it making an angle
$\theta (\theta \neq 0)$ with stick. If the stick falls without slipping the length of its shadow will increase
at first and then decrease. The max. length of the shadow will be

a) L

b) $\frac{L}{cos\theta}$

c) $L\ tan\theta$

d) $L\ cos\theta$

#11 If transparent liquid is poured into the concave part of a converging meniscus lens

a) The focal length will increase.
b) The focal length will decrease.
c) The focal length will not change.
d) The change of focal length deponds on the index of refraction of
the lens and the liquid.

#12 We connect the positive terminal of a battery to an aluminium pressure cooker and its
negative terminal to a kitchen knife. Which of the following statements is correct?

a) The magnitude of charge on pressure cooker is more than its magnitude on kitchen
knife.
b) The magnitude of charge on the kitchen knife is more than its magnitude on pressure
cooker.
c) Their total charge is zero.

d) All the above statements may be correct under special condition.

#13 It is intended to increase the capacitance of a parallel plate capacitor, which are 1 mm apart as much as possible. If we insert a piece of material 0.9 mm thick, which of the substance below fulfil our requirement.

a) Aluminium   b) Mica   c) Hard plastic   d) Glass

#14 Two similar conducting balls having positive charges q_1 and q_2 are separated by distance r. If they are made to touch each other and then separated again to the same distance, the force in this case will be

a) Less than before   b) More than before

c) Same as before   d) Zero

#15 Eight similar batteries are connected by wires of negligible resistance as shown in Fig(3). If the emf. of each battery is 5 volts and its internal resistance is .2 Ω the voltmeter connected between points A and B will read

a) Between five and forty volts.
b) Between zero and five volts.
c) Zero
d) Thirty five volts.

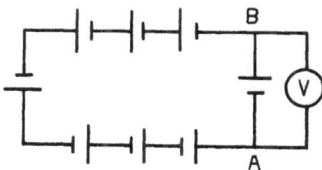

Fig(3)

Correct answers of the mutiple choice questions

1	c	6	c	11	b
2	b	7	c	12	c
3	b	8	b	13	a
4	c	9	c	14	b
5	b	10	b	15	c

II-Some of the problems and their solution given in the second stage of the preliminary competition, in June 1989.

1- a) Why the diameter of a continuously falling stream of water from an open faucet decreases as it falls?

b) Suppose a faucet has a diameter of 1 cm and is located at the heigth of 75 cm. Determine the diameter of falling stream of water from it at the ground level, if the speed of water at the faucet is $1m/s (g = 10m/s^2)$
Solution:

a) The volume of water crossing in time t, at any cross section of stream, is constant. Since the velocity at the top of the stream is less than the velocity at the lower part of it, therefore the diameter of the stream should be greater at the upper part than the lower part.

b) $v^2 - v_0^2 = 2gh$

$v^2 = 1^2 + 2 \times 10 \times 0.75 = 16$ $\qquad v = 4 \ cm/s$

$D_0^2 v_0 = D^2 v$ $\qquad D = D_0 \frac{\sqrt{v_0}}{\sqrt{v}} = 1 \times \sqrt{1/4} = 0.5 \ cm$

2- An inverted U shaped rod without electrical resistance is placed vertically on the ground.(see the fig 4)
Another rod with length L,
mass m and electrical resistance
R, with the aids of two frictionless rings is sliding downward
on it.A homogenous magnetic field B,
perpendicular to the plane of
inverted U shape rod is also present.

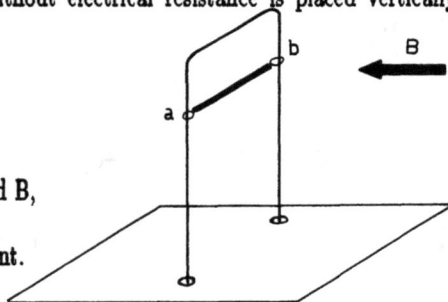

Fig (4)

a) If the sliding rod begins to
fall, under the influence of its weight, after certain time it acquires a constant velocity known as terminal velocity. Explain (at most in 5 lines) why such a velocity is attained by the rod.

b) Having acquired terminal velocity, calculate the magnitude and direction of induced current in the rod, and its terminal velocity.

c) Suppose the rod moves the distance h with terminal velocity in time t. By calculation show that the gravitational potential energy lost by the rod, is equal to the heat dissipated in it.

Solution:

a) As the downward movement starts, there will be an induced current in the rod. Therefore a Lorentz force is exerted on the rod which is, by the Lenz's law, opposite to the weigth of the rod. As a result the acceleration of the rod gradually tends to zero. At that instant the velocity known as terminal velocity, will stay constant,

b) $F = iLB = mg \Longrightarrow i = \frac{mg}{LB}$

$E = iR = BLv$

$v = \frac{iR}{BL} = \frac{\frac{mg}{LB}R}{BL} = \frac{Rmg}{B^2L^3}$

By applying right hand rule the direction of the induced current is determined. It is shown in fig(5).

c) $h = vt$ $\qquad \Delta U = mgh = mgvt$

$\Delta Q = Ri^2t = R\frac{mg}{B^2L^2}mgt = \Delta U$

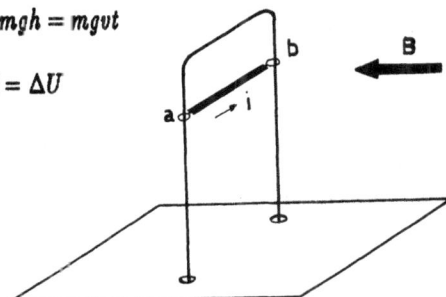

Fig(5)

3- Suppose two capacitors with capacitances c_1 & c_2 and charges q_1 & q_2 respectively are connected so that the plates with similar charges are connected to each other.

a) Using physical reasoning and without calculation explain (at most in 6 lines) what changes will occur in the energy of the system with respect to initial energy of the capacitors.

b) By calculation derive the result you have obtained in part (a).
Solution:

a) If the potential differences of capacitors are not the same, there will be a potential difference between the positive plates, as we connect the negative plates together. Now as the positive plates are connected positive charges will move from higher potential to the lower potential and the amount of (qV) energy is converted into heat, making final energy of the system less than the initial energies of the capacitors.

b) $U_i = \frac{q_1^2}{2c_1} + \frac{q_2^2}{2c_2}$ $\qquad\qquad\qquad U_f = \frac{(q_1+q_2)^2}{2(c_1+c_2)}$

$$2(U_i - U_f) = \frac{q_1^2}{c_1} + \frac{q_2^2}{c_2} - \frac{q_1^2 + q_2^2 + 2q_1q_2}{c_1 + c_2} =$$

$$\frac{c_2(c_1 + c_2)q_1^2 + c_1(c_1 + c_2)q_2^2 - c_1c_2q_1^2 - c_1c_2q_2^2 - 2c_1c_2q_1q_2}{c_1c_2(c_1 + c_2)} =$$

$$\frac{c_2^2q_1^2 + c_1^2q_2^2 - 2c_1c_2q_1q_2}{c_1c_2(c_1 + c_2)} = \frac{c_1c_2}{c_1 + c_2}\left(\frac{q_1}{c_1} - \frac{q_2}{c_2}\right)^2 =$$

$$\frac{c_1c_2}{c_1 + c_2}(v_1 - v_2)^2$$

If $v_1 \neq v_2$, always part of the energy is lost in the form of heat.

III-Some of the problems and their solution given in the semifinal stage (for selecting 7 student out of 30) in Feb.1989

1- Consider a bicycle where each
wheel has a radius R and a mass
m(assume the masses are distributed
uniformly on the
circumferences of the wheels).
If the total mass of the bicycle
and the bicyclist is M
and the center of mass is at
height h and distance l

Fig(6)

(the distance from vertical line through center of mass and place where
the wheel touches the ground), as shown in the fig(6)
By assuming the torque due to paddling on the rear wheel to be τ.
(a) Find bicycle's acceleration,
(b) Find the max.value of τ so that the wheel will not slip (coefficient of static friction $= \mu_s$)
(c) If μ_s is large enough so that there is no slipping, is there any max valur for τ ?

Solution:
(a) From forces in horizontal
direction and
torques in direction
perpendicular to the
page,we have (see fig 7)

$$f_s - f_{s'} = ma$$

$$\tau = f_s R = I\alpha$$

and

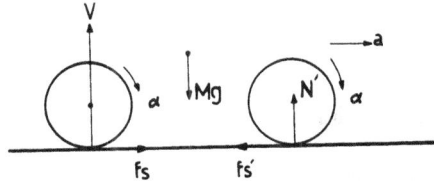

$$f_{s'} R = I\alpha$$

Fig(7)

Using equation of constrain and $I = mR^2$, we obtain

$$f_s - f_{s'} = ma \qquad\qquad \tau/R - f_s = ma$$

$$\tau/R = (m + 2m)a \qquad a = \frac{\tau/R}{M+2m}$$

$$f_{s'} = ma$$

(b) in this part we must determine the forces N and N'. since $mg = N + N'$ and $\tau_{ext.} = dL/dt$,we must first calculate the torques and total angular momentum with respect to the center of mass system. The system consists of three parts, front wheel,bicyclist,rear wheel and the body of bicycle. The angular momentum of these three parts are the same as angular momentum of the center of mass (since they have same translational velocity),

$$\mathbf{L} = m_1\mathbf{r}_1 \times \mathbf{v} + m_2\mathbf{r}_2 \times \mathbf{v} + m_3\mathbf{r}_3 \times \mathbf{v} = (m_1\mathbf{r}_1 + m_2\mathbf{r}_2 + m_3\mathbf{r}_3) \times \mathbf{v} = m\mathbf{R}_{cm} \times \mathbf{v}$$

in center of mass system the total angular momentum is zero,and so it is sufficient if we calculate angular momentum of each part about its own center of mass. If we take positive direction into the page,we have

$$L = I\omega = 2mR^2 v/R = 2mRv$$
$$\tau_{ext.} = Nl - f_s h - N'l + f_{s'}h$$

and
$$(N - N')l + (f_{s'} - f_s)h = dL/dt = 2mRa$$
From (a), $f_s - f_{s'} = ma$

$$N - N' = \frac{h}{l}Ma + \frac{2R}{l}ma = \frac{hM+2Rm}{l}a = \frac{hM+2Rm}{l} \times \frac{\tau/R}{M+2m}, N + N' = Mg$$

$N = 1/2(Mg + \frac{hM+2Rm}{M+2m}\tau/lR)$ and $N' = 1/2(Mg - \frac{hm+2Rm}{M+2m}\tau/lR)$

we also have, from part (a);

$f_{r'} = ma = \frac{m}{M+2m}\tau/R$

and

$f_r = f_{r'} + Ma = (m+M)a = \frac{M+m}{M+2m}\tau/R$

The no slip conditions are

$f_{r'} \leq \mu_s N'$ and $f_r \leq \mu_s N$

thus

$\frac{M+m}{M+2m}\tau/R \leq 1/2\mu_s(Mg + \frac{hM+2Rm}{M+2m}\tau/lR)$

$\tau \leq MgR\frac{\mu_s(M+2m)}{(2-\mu_s\frac{h}{l})}\mu + (2 - 2\mu_s\frac{R}{l})m$

$\frac{m}{M+2m}\tau/R < 1/2\mu_s(Mg - \frac{hM+2Rm}{(M+2m)}\tau/lR)$

$\tau \leq MgR\frac{\mu_s(M+2m)}{\mu_s\frac{h}{l}M+(2+2\mu_s\frac{R}{l})m} = \tau_{2m}$

If τ is greater than τ_{1m} or τ_{2m} the front or rear wheel will slip, τ must be smaller than the minimum of τ_{1m} and τ_{2m}. If $\tau_{2m} \leq \tau_{1m}$,first the front wheel will slip,the condition for this will be

$(2 + 2\mu_s\frac{R}{l})m + \mu_s\frac{h}{l}M \geq (2 - \mu_s\frac{h}{l})M + (2 - 2\mu_s\frac{R}{l})m$

or

$\mu_s > \frac{Ml}{2Rm+hM}$

Otherwise the rear wheel will slip. Note in these cases the front wheel will slide.
(c) If the friction is great enough so that increasing τ will not cause the wheel to slip, still τ can not be increased as much as we want, because by increasing τ,N' approaches zero and all the weight falls on the rear wheel. Note if τ increases further the front wheel comes up, and the system rotates around the center of mass. i.e.the bicyclist falls on his back,we must have,

$N' > 0, 1/2(Mg - \frac{hM+2Rm}{M+2m}\tau/lR) > 0 \implies \tau \leq MgR\frac{(M+2m)l}{hM+2RM}$

2- A satellite with mass m is rotating in a circular orbit of radius r_0 around the earth. If the mass of earth is taken as M,

(a) Determine the total mechanical energy of satellite?

(b) Suppose the satellite rotates in the earth upper atmosphere where there is a constant frictional force f. The satellite will move in a spiral toward the earth and since, f is a weak force the change in the radius is so small that one can assume at each moment the orbit is a circle with an average radius r.

Find approximate change in the radius, Δr, for each revolution.

(c) Calculate the approximate change in the kinetic energy of the satellite for each revolution.

Solution:

$$K.E. = (1/2)mv^2 \qquad P.E = -G\frac{mM}{r_0}$$

(a) From Newton's 2nd law,

$$\frac{GmM}{r^2} = \frac{mv^2}{r_0} \implies mv^2 = \frac{GmM}{r_0} \implies K.E = (1/2)\frac{GmM}{r_0}$$
$$E = K.E + P.E. = -(1/2)\frac{GmM}{r_0}$$

(b) Because of frictional force, mechanical energy is not conserved and since the rate of change of energy is equal to power dissipated by frictional force,

$$dE/dt = \mathbf{f}.\mathbf{v} = -fv$$
$$d/dt(-1/2\,GmM\frac{1}{r}) = -fv(r) = -f(GM/r)^{1/2} + \frac{GmM}{2}\frac{\dot{r}}{r^2} = -f(GM/r)^{1/2} \implies \dot{r} = -\frac{2f}{m(GM)^{1/2}}r^{3/2}$$

$$\frac{\Delta r}{\Delta t} = \dot{r} \implies \Delta r = -\frac{2fT}{m(GM)^{1/2}}r^{3/2}T, \text{if } \Delta t \text{ is one period}$$

The negative sign indicates the radius is decreasing,

(c) $K.E. = 1/2\frac{GmM}{r} = -E$

$$\frac{d}{dt}(K.E.) = -dE/dt = -\frac{GmM}{2r^2}\dot{r} = \frac{GmM}{2r^2}\frac{2f}{m(GM)^{1/2}}r^{3/2}$$
$$\frac{d}{dt}(K.E) = f(GM/r)^{1/2} \implies \Delta K.E. = f(GMm/r)^{1/2}T, \text{the kinetic energy is increasing.}$$

3- A solid sphere with radius r and mass m rotates inside a hollow sphere of radius R, as shown in the fig(8). Calculate a) period of small oscillation about equilibrium, and b) frictional force when θ is maximum. (moment of inertia for solid sphere is $2/5mr^2$).

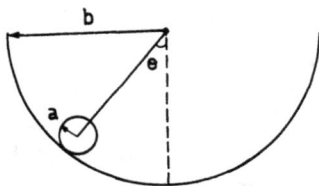

Fig(8)

Solution:

a) Let θ be the angle of small sphere with respect to vertical line and ψ be its angle of rotation about its own axis, see fig(9).

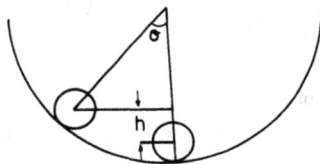

Fig(9)

$(R - r)\theta = r\psi$
and
$(R - r)\dot{\theta} = r\dot{\psi} \implies (R - r)\ddot{\theta} = r\ddot{\psi}$

The kinetic energy of the small sphere is .

$$K.E. = 1/2 m v_{2cm} + 1/2 I \dot{\psi}^2 = 1/2 m (R - r)\dot{\theta}^2 + 1/2 (\tfrac{2}{5} m r^2)(\tfrac{R-r}{r}\dot{\theta})^2$$
or

$$K.E. = \tfrac{7}{10} m (R - r)^2 \dot{\theta}^2$$

If we take the potential energy to be zero when $\theta = 0$, we have

$$P.E. = mgh = mg(R - r)(l - \cos\theta),$$

$$E = K.E. + P.E. = mg(R - r)(l - \cos\theta) + \tfrac{7}{10} m (R - r)^2 \dot{\theta}^2$$

thus

$$dE/dt = 0 \implies mg(R-r)\dot{\theta}\sin\theta + \tfrac{7}{10}m(R-r)^2 2\dot{\theta}\ddot{\theta} = 0$$

or

$$g\sin\theta + \tfrac{7}{5}(R-r)\ddot{\theta} = 0$$

For small oscillation we use the approximation $\sin\theta \simeq \theta$ in the above equation, we obtain

$$\ddot{\theta} + \tfrac{5}{7}\tfrac{g}{R-r}\theta = 0 \implies \omega = 5g/7(R-r)^{1/2}$$
or

$$T = 2\pi/\omega = 2\pi 7(R-r)/5g^{1/2}$$

b) To calculate the frictional force at some angle θ note, the frictional force is the only force which has a torque about the center of mass (see fig 10). therefore

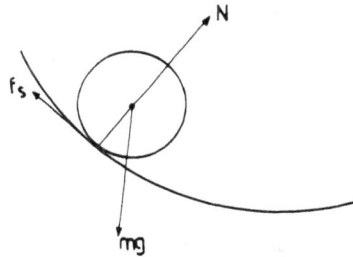

Fig(10)

$$f_s r = I\,|\,\ddot{\psi}\,|, \quad f_s r = \tfrac{2}{5}mr^2\tfrac{R-r}{r}\,|\,\ddot{\theta}\,|,$$

$$f_s = \tfrac{2}{5}m(R-r)\ddot{\theta} = \tfrac{2}{5}m(R-r)\tfrac{5}{7}\tfrac{g}{R-r}\,|\,\theta\,|,$$

and

$$f_s = \tfrac{2}{7}mg\,|\,\theta\,|$$
which for max.θ, i.e.θ_{max}, we have
$$f_s = \tfrac{2}{7}mg\,|\,\theta_{max}.\,|$$

4- A small solid sphere of mass m and radius r approaches another larger solid sphere of mass M and radius R with velocity v_0. Take origin on the center of larger sphere and the z-axis parallel to v_0 such that the distance from center of small sphere to z-axis (see fig 11) is b.

(a) Find the final velocity of the small sphere with appropriate physical assumption (explain your assumption clearly).

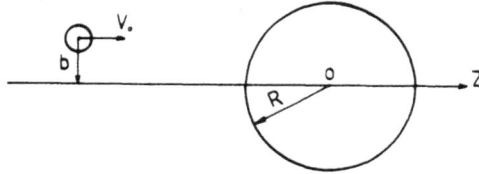

Fig(11)

(b) A thin layer of gold with thickness d is bombarded by a beam of alpha particles with velocity v_0. If incident alpha particles and gold atoms are taken as the above spheres respectively and number of alpha particles per unit time per unit area perpendicular to the beam is I_0, and the density of gold atoms(number of atoms per unit volume) is n, find the number of alpha particles per unit time which will be scattered between angle θ and $\theta + d\theta$ with respect to z-axis (direction of incident beam).

(c) With the assumptions used in part (b), about gold atoms and alpha particles, what important predictions can be concluded about scattering of alpha particles? If these predictions come true what physical quantities can be measured?

Solution:

(a) Assume the force acting on the small sphere upon contact is along the line joining the centers of two spheres,when they are at contact.The momentum of the small sphere will not change in the direction perpendicular to this line, however since some of the energy of the small sphere will be given to the larger one,its momentum will be smaller after the collision.

Assume we have an elastic collision,using conservation of momentum in the direction perpendicular to the z-axis we have: (see fig 12)

$$mv\sin\theta - MV\sin\psi = 0 \Longrightarrow \frac{V}{v} = \frac{m\sin\theta}{M\sin\psi}$$

and

$$\frac{1/2MV^2}{1/2mv^2} = \frac{m}{M}\frac{\sin^2\theta}{\sin^2\psi}$$

So the ratio of the kinetic energy of the large sphere to the small one goes as m/M. To calculate the final velocity, we can simplify the calculation if we assume $\frac{m}{M} \ll 1$, then the kinetic energy of the larger sphere may be neglected. The conservation of energy will demand that the final momentum of small sphere will be the same as its initial value i.e. the small sphere will be scattered with the same angle as incident, $\psi = \psi'$.

$$\theta + 2\psi = \pi \Longrightarrow \tfrac{\theta}{2} + \psi = \tfrac{\pi}{2} \longrightarrow \cos\theta/2 = \sin\psi$$

$$b = (R+r)\sin\psi = (R+r)\cos\theta/2$$

By knowing b we can calculate θ.Note the velocity of incident particle will not change, and the velocity of larger sphere is not appreciable.

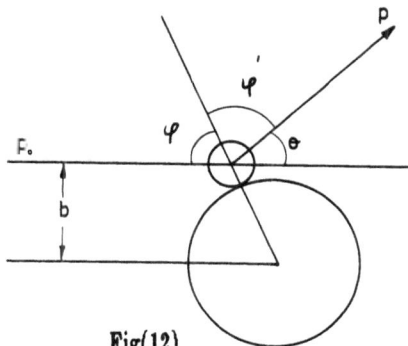

Fig(12)

(b) Let's calculate the cross section for the particle to be scattered between angles θ and $\theta + d\theta$.(see fig 13)
If small sphere covers the area πb^2,

$$\sigma(\theta) = 2\pi b d\bar{o}$$

Using $b = (R+r)\cos\tfrac{\theta}{2}$

$$db = (R+r)(-1/2)\sin\tfrac{\theta}{2}d\theta,$$

or

$$|\,db\,| = \tfrac{R+r}{2}\sin(\tfrac{\theta}{2})\,|\,d\theta\,|,$$

thus
$$\sigma(\theta) = 2\pi b\tfrac{R+r}{2}\sin(\tfrac{\theta}{2})\,|\,d\theta\,| = \pi(R+r)^2\cos(\tfrac{\theta}{2})\sin(\tfrac{\theta}{2})\,|\,d\theta\,|$$

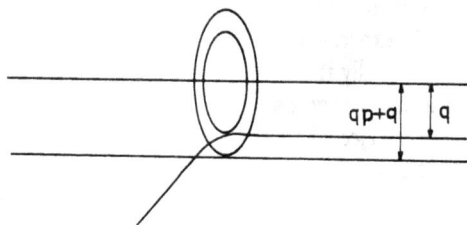

or

$$\sigma(\theta) = \tfrac{\pi}{2}(R+r)^2\sin\theta\,|\,d\theta\,|$$

Fig(13)

For each gold particle we should consider such a scattering cross section. Consider a narrow beam of particles incident upon the unit cross section,and assume the number of particles incident on this area per sec,to be I_0 (see fig 14). There are $d \times 1 \times n$ number of gold atoms per unit area in the thin layer of gold and therefore the effective cross section will be equal to this number times the scattering cross section $\sigma(\theta)$; i.e. ;

$$I(\theta) = I_0 dn\sigma(\theta)$$

or

$$I(\theta) = I_0 dn\tfrac{\pi}{2}(R + r)\sin\theta \mid d\theta \mid$$

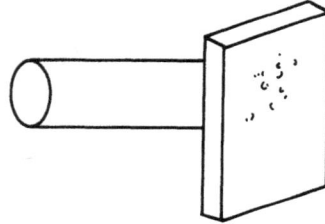

Fig(14)

Here the approximation we used in part (a) is appropriate since the gold atoms are not free particles,and practically we should consider a larger mass for them , than their own atomic masses.

(c) If we draw $I(\theta)$ versus θ,using the above formulas(fig 15),we see that the number of α particles scattered in forward direction should be the same as those ejected back. However this is in contradiction with experiment, since experimentally it has been shown that a very few particles will be ejected back and the scattering of α particles are proportional to $\frac{1}{\sin^4(\frac{\theta}{2})}$ which is the same

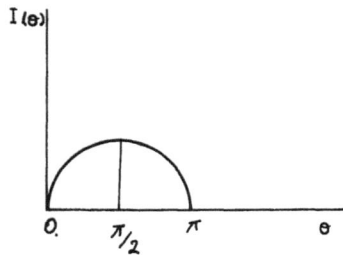

Fig(15)

as Rutherford scattering theory. If the assumption we used is confirmed by the experiment we could measure the size of gold atom nucleus (as solid sphere). If the dimension of α particles were known or if we assume they are very small with respect to gold atoms, by exact measurement of $I(\theta)$, we can calculate the radius of gold atom nucleus.

ITALY

PROCEDURES FOR SELECTING ITALIAN TEAMS TO THE INTERNATIONAL PHYSICS OLYMPIADS

P. Violino
(Torino, Dipartimento di Fisica dell'Universita')
G. Cavaggioni
(Venezia, Liceo Scientifico G. Bruno)

Every year the Ministry of Public Instruction decides the participation of the Italian team at the International Olympiads of Physics and communicates the fact to all secondary schools. Since 1987 all the activities concerned with the selection and training of the Italian contingent have been entrusted to the Associazione per l'Insegnamento della Fisica - A.I.F. (Association for the Teaching of Physics). A National Commitee of A.I.F. made up of professors and secondary school teachers plans and coordinates the activities with the assistance of local groups.

The competions are organized keeping the following aims in mind: maintaining and stimulating further the interest of the students for the study of Physics; promoting and innovating the teaching of Physics in terms of both contents and method; allowing the more able and motivated students to find the opportunity of carrying out more in-depth studies than normally possible in their regular school programmes.

The secondary schools which would like their students to participate in the competitions request as much without restrictions. Physics is usually taught either in the first two years of a five-year school (technical colleges) or in the final two or three years (licei- Grammar School-type colleges); currently in some experimental-type colleges Physics is taken for five years. A selection of physics problems of different grades and difficulties is sent to the schools which have asked to take part in the competitions along with all the problems used in the previous years' national competitions and International Olympiads. This material can be used by teachers for the special preparation of those students who will take part in the competition.

Participating students are required to have already

followed at least one annual course of Physics.Each school decides which students will participate in the first series of tests: it is recommended to allow all students interested in Physics to take part in this; the first phase of selection is carried out within each school. In general all interested students can participate although each school decides autonomously which of its students take part in this first internal selection.

The first round of tests consists of a closed-answer set of questions, which is the same for all schools and which is given to the students on the same day under the same conditions. These questions are formulated by the National Committee. Since the test is the same for students of different levels, the content of Physics and difficulty and number of the questions are so organized as to allow also the more able students of lower levels to obtain a good score. The scores are assigned by the teachers by using a grid-scheme of marking which is supplied along with the set of questions: on the basis of the results each school can pick out a maximum of five students to take part in the second phase of the competition. The National Committee receives the total results regarding the answers to the tests and the answer sheets of the nominated students. In 1989 the papers of 3000 students were sent in, 393 of which were finally indicated.

The second stage is organized in collaboration with local groups, usually on a regional basis. In this second competition the solutions to theoretical problems are asked for. The National Committee prepares the problems which are the same for all competitors along with the evaluation grid. The correction of the problems is made by local groups. The competition takes place on the same day in all the test centres. There are four or five problems dealing with numerous subject-matters of Physics, distributed so as to cover the programmes of all the levels. The problems are not as difficult as those of the Olympiads.

Before this second stage the local groups can organize preparative courses for the students, meetings with teachers and, after the competition, can award prizes to the best students of their area, even if they do not appear in the national classification of those admitted to the third stage. The finances of these activities are raised by the local groups.

The 50 students who obtained the best scores in the Regional Competitions are invited to participate in the National Competition. The National Competition has so far been held at Viareggio, a seaside resort on the coast of Tuscany, not far from Pisa. The accepted policy for this stage is to set theoretical problems, usually four, to be solved in four hours; the difficulty of these is comparable to those at the Olympiads. Marking is done by the National Group which assigns the scores according to the pre-established evaluation grid. Ten winners are announced, whose names are immediately communicated to their respective schools.

The subject-matters of the three tests are taken from the Italian Physics programmes and part of those expected in the syllabus of the Olympiads is not covered. It is hoped that the diffusion of the problems posed at the Olympiads and the articles of comments on the contents and resolutive methods, achieved through the A.I.F.'s review, can hasten the growth and innovation of official Physics programmes. It is only in this way that the bases of the subject-matters we are lacking in today can be widely taught in schools with courses appropriately organized and well staged over time. The lackings of the current programmes damage less the students who are the best and most interested in the study of Physics and who can easily make up for such deficiencies in their university courses. Instead, the holes left in the general formation of basic scientific culture are irreparable for those who do not further such studies.

The ten winners of the National Competition are invited to follow an intensive training course lasting a week at the University of Bologna. The course is planned by the National Committee together with a group of professors from the host university. It comprises some sessions of problem-solving but most of the time is devoted to improving the students' experimental abilities. The use of the laboratory in the teaching of Physics is not compulsory in the Italian licei and in general the preparation of the students in this field is highly unequal. During the training week the students undergo certain tests and, on the basis of the results, the team is selected.

No scolastic bonuses are currently expected for the competition winners; they take all their final secondary school exams, including Physics, as normal, and are not granted any concessionary treatment when entering university.

THEORETICAL PROBLEM
(Italian National Competition 1988)

Paolo Nesti (Livorno, Liceo Sperimentale Cecioni)

A small metal ball of mass m can slip, threaded on a smooth, vertical circular wire of radius R. The ball is first displaced at the point P_0 such that $P_0OA = 60°$, and then released. (Fig. 1)

DATA: m = 1 kg ; k = 50 N/m ; R = 50 cm .

Neglect the mass of the spring and the friction between the ball and the wire.

1) Derive the condition the given quantities must fulfil assuming the ball can reach the point B. Verify that given data satisfy the condition above.

2) Find the equilibrium positions of the ball along the wire. Is B a stable equilibrium point ?

3) Find the speed of the ball when it reaches the point B.

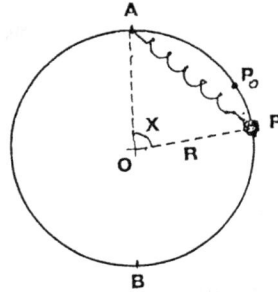

Figure 1

4) Another, identical, little ball is now at rest at the point B; it can slip too along the wire with negligible friction. Determine the highest points the two balls can reach if they collided elastically.

SOLUTION

1) Three forces can act on the ball at any point P of the wire: its weight, the elastic force due to the spring and the reaction of the wire. Since $AP_0 = R$ the spring is unstretched and no elastic force is initially acting on the ball.

Neglecting the friction we can use conservation of energy:

(1) $[E_g]_{P_0} = [E_g]_P + [E_{e1}]_P + [E_k]_P$

where $[E_g]_P$ is the gravitational energy, $[E_{e1}]_P$ the elastic energy and $[E_k]_P$ the kinetic energy of the ball at any point P of the wire.

Notice that, at P_0, $E_k = E_{e1} = 0$.

The ball will reach the point B only if:

(2) $[E_k]_B = [E_g]_{P_0} - [E_g]_B - [E_{e1}]_B \geq 0$.

since $(E_g)_P - (E_g)_P$ and $(E_{e1})_P$ are increasing functions of the point P when the ball goes from P_0 to B.

We may calculate:

(3) $[E_g]_{P_0} - [E_g]_B = 3/2\ R\ m\ g$

(4) $[E_{e1}]_B = 1/2\ k\ R^2$

substituting these values into the (2),

(5) $$m \geq \frac{k\ R}{3\ g}$$

which is the required condition.

It can be observed that given data satisfy this condition as

$$\frac{k\ R}{3\ g} = \frac{50\ N/m\ 0.5\ m}{3 \cdot 10\ m/s^2} = 0,83\ kg$$

and the mass of the ball is m = 1 kg.

2) Equilibrium.

The ball will be in equilibrium at a point P if its weight and the force exerted by the spring have equal components along the tangent to the circle at P. (Fig. 2)

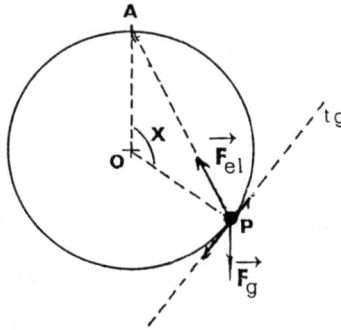

Figure 2

(6) $F_{el} \cos(x/2) = F_g \sin x$ where x = AOP .

Since

AP = 2 R sin(x/2) ; F_{el} = k [2 R sin(x/2) - R] ;

F_g = m g ;   sin x = 2 sin(x/2) cos(x/2);

substituting in equation (6), we have the (7):

kr [2 sin(x/2) - 1] cos(x/2) - 2 mg sin(x/2) cos(x/2) = 0 .

   This equation is true if

(8)   cos (x/2) = 0

or

(9) $$\sin (x/2) = \frac{k\ R}{2\ (k\ R - m\ g)}$$.

   We may observe that, according to the (8), the point B is always an equilibrium position. The equation (9) implicates that 2(k R - m g) > k R , since sin(x/2) > 0 ; hence

(10) $$m < \frac{k\ R}{2\ g}$$.

   With the given data, kR/2g = 1.25 kg,   two more equilibrium positions are allowed:

$$x_1 = 180° - 67° = 113°$$

$$x_2 = 180° + 67° = 247° \quad .$$

Equilibrium at B.

When the ball is slightly deflected from B it is pushed back by its weight and pulled farther from B by the spring. The condition for a stable equilibrium is:

(11) $(F_g)_{tg} > (F_{el})_{tg}$

taking into account the tangential component of the forces. (Fig.3).

So, if $a = OAP$, we have,

(12) $mg \sin 2a > k R \sin a$.

Notice that, close to the point B, the extension of the spring is almost R. In this case we have also $a \approx 0$, $\sin a \approx a$ and $\cos a \approx 1$.

The (12) becomes

(13) $2 m g a > k R a$;

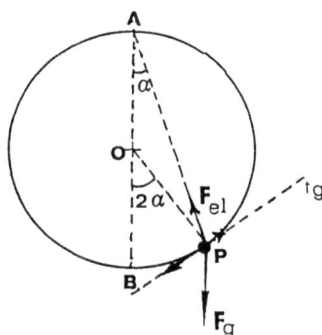

Figure 3

hence B is a stable equilibrium position if

(14) $m > \dfrac{k R}{2 g}$.

So, according to condition (10), with three equilibrium positions, the equilibrium at B is unstable.

3) Using conservation of energy, when the ball is at the point B we have:

$$[E_k]_B = [E_{\varepsilon}]_{P_0} - [E_g]_B - [E_{el}]_B$$

(15) $1/2 \ m \ (v_B)^2 = 3/2 \ m \ g \ R - 1/2 \ k \ R^2$

hence

$$v_B = \sqrt{3 \ R \ g - (k/m) \ R^2} \qquad .$$

With the given data, $v_B \approx 1,58$ m/s .

4) Since the balls impinge directly, after the impact the striking ball is brought to rest and the second one begins to slip along the wire with the same speed v_B of the first ball at the impact.

4a) The pushed ball slips and comes to rest at its highest point P_1 ; according to conservation of energy:

(16) $$[E_g]_{P_1} - [E_g]_B = [E_k]_B$$

Taking B as the level of zero gravitational energy and $x_1 = AOP_1$

(17) $$[E_g]_{P_1} = m \, g \, R \, (1 + \cos x_1) \, .$$

From (15)

(18) $$[E_k]_B = 1/2 \, m \, [3 \, R \, g - (k/m) \, R^2] \quad .$$

Substituting the (17) and (18) into the (16)

$$\cos x_1 = - \frac{k \, R}{2 \, m \, g} = - 3/4$$

$$x_1 = 221° \quad \text{clockwise from A.}$$

4b) The pushing ball is unlikely to stay at rest in an unstable equilibrium position. We can find the highest point P_2 that it can reach after having been slightly displaced from B using conservation of energy:

(19) $$[E_g]_{P_2} - [E_g]_B = [E_{el}]_B - [E_{el}]_{P_2}$$

Substituting the following equations into the (19)

$$[E_g]_{P_2} - [E_g]_B = m \, g \, R \, (1 + \cos x_2)$$

$$[E_{el}]_{P_2} = 1/2 \, k \, R^2 \, [2 \, \sin \, (x/2) - 1]^2$$

$$[E_{el}]_B = 1/2 \, k \, R^2$$

we have

(20) $(k \, R - m \, g) \, \sin^2 \, (x_2/2) - k \, R \, \sin \, (x_2/2\cdot) + m \, g = 0$

which gives the solutions:

(21) $\sin (x_2/2) = 1$ for the ball resting at B.

(22) $\sin (x_2/2) = \dfrac{m \, g}{k \, R - m \, g}$.

The (22) gives always solutions when the unstable equilibrium condition, $m < (kR)/(2g)$, is fulfilled. With the given data; $\sin (x_2/2) = 2/3$ and $x_2 = 82.5°$. The angle is taken clockwise or counter-clockwise from A according to the displacement of the sphere when it leaves the unstable equilibrium point, B .

THE NETHERLANDS

BRIEF HISTORY OF THE PHYSICS OLYMPIAD IN THE NETHERLANDS

Since 1982 the Netherlands participate in the international physics olympiad and hence the national physics olympiad has been organised since 1982 as well.

The Netherlands being invited to the international physics olympiad in the FRG the Ministry of Education gave notice of this invitation by the end of March 1982. The institute for teacher training of the University of Groningen accepted the invitation and started to organise a national contest to ensure the participation in the international olympiad. By doing so 32 students of five different schools took the test. For reasons of convenience the test was a theoretical test with mainly multiple choice questions and only two open questions.

As a result of this test twelve highschool students were invited to take part in the second stage of the national olympiad. This second stage consisted only of one experimental test which was held at the University of Utrecht. No additional lectures or tests were given and only the result of the primer theoretical test and the experimental test determined who would to be choosen as a participant of the international physics olympiad.

So the very first Dutch team that ever went to an international physics olympiad was rather unprepared. Nevertheless the results were promising. Therefore the Dutch organisers decided to continue the organisation of a national physics olympiad to ensure the participation in international olympiads. Our first concern was to raise funds to make the organisation of national and international olympiads possible. The University of Groningen, where the organisation took place, was very helpful. It not only provided money but also gave permission to two of its members to regard the organisation of the Dutch physics olympiad as part of their job. Apart from the University of Groningen also the Ministery of Education financed the olympiad for a substantial part. For the rest some industries did contribute.

So from the very start the financial part of the organisation seemed to be settled. What was left was to ensure a sound organisation that would be able to select a team that could challenge the standards of the international physics olympiad. That task was less easier than it appeared to be. Although we learned a lot from our first experience in 1982 in the FRG we only came to the final form of our organisation by the years. The experience during the international olympiads was very helpful but not decisive. The first major problem faced by the Dutch physics olympiad was the olympiad being accepted in the Netherlands as a contest that was mainly organised for the very good students in physics.

During the seventies a large educational effort was put in the development of the skills of those boys and girls in highschool who were poor in cognitive matters. This effort was highly supported by the Ministery of Education and in fact was part of the policy of the socialist gouvernments ruling the Netherlands in those days. The idea that the gifted didn't need any help or special attention at school was widespread. All the attention was directed towards those who faced more or less difficulties in their highschool careers as well in social as in cognitive respect. Therefore the acceptance of a contest for the best in physics was sometimes a difficult task to fulfil especially when the cooperation of the highschool teachers was needed.

In the first stage of the contest the theoretical problems are sent to the schools who are willing to participate and the teachers are asked to be so kind to correct the open questions according to a model. Fortunately more and more teachers do like to do so and an increasing number of schools and students are participating in the physics olympiad. Two major reasons are responsible for the success of the physics olympiad in the Netherlands. First of all the organisers try to select problems that are challenging for everybody who takes physics courses in highschool. The main purpose for the physics olympiad is therefore not to select a national team for the international olympiad, but to present a selection of problems that is 'worth to be solved'. In this way we not only motivate students to participate but also the teachers who can use the problems for their own teaching although mostly in a modified way.

A second reason for the succes we are facing now is the changing attitude towards contests as a whole and the physics olympiad in particular. There are schools in the Netherlands where the physics teacher is even able to organise a school contest in physics and where the national olympiad serves as such. In those cases all the students in the final classes participate whether they have good results in physics or not. Such schools even provide prizes on their own account for their students.

ORGANISATION OF THE NATIONAL PHYSICS OLYMPIAD.

Starting on a small basis the dutch physics olympiad has grown out to a mature organisation that runs the national contest with a minimum of staff and a modest budget. Members of the scientific community of the University of Groningen and Utrecht and a representative of the dutch physics teachers organisation form the heart of the organisation. With a budget that mounts up to 40.000 Dfl (in 1989) we are able to organise a national contest in which over a thousand students participate. That number is still increasing with about 20% every year.

The contest consists of two grades. During the first grade, which takes place at the end of January, the students have to solve theoretical problems. The test is sent to their school about a week in advance. Teachers and staff of the school have been notified of this test far ahead so they will be able to take all necessary arrangements to organise the contest on the appropiate day and hour. Only schools that apply for the national olympiad receive the tests. Every year complete information about the olympiad is given by the Ministery of Education to the administration of all secondary schools in the Netherlands. It is left to the initiative of the physics teachers to promote the participation of their students in the physics olympiad. Individual students cannot apply for the contest; they have to do that by the intermediate of their teachers. When subscription of the students is received by the physics olympiad organisation, copies of the problems for the first round are sent to the teachers, and they are asked to organise the first contest, which is a theoretical test, on a fixed date and under their supervision. The test will take three hours and consists of 30 multiple choice questions and 6 open answer questions which cover not only the physics taught in highschool, but may contain as well problems about 'unknown' subjects. Any specific knowledge about these 'unknown' subjects is not requested. The papers of the students are corrected by the teachers, according to an answering model, before they are sent back to the physics olympiad organisation. The answers for the multiple choice questions and the scores for the open answer questions are checkued by the authors of the test and by computer, and in this way an order in the total scores for the first round is produced. The 20 highest scoring students are admitted to participate in the second round of the olympiad. Therefore they are invited at one of the organising universities in the Netherlands (Groningen and Utrecht) to be their guest for one week during which they follow lectures and participate in laboratory work. Two tests will be held as well; a theoretical one and an experimental one. The scores for these tests together with the score in the first round determine the final order in the total scores of the students. All 20 students are awarded in a ceremony in which a special prize is given to the highest scoring girl. In general the Minister of Education or his Secretary of State leads the awarding. The five best students form the team that will participate in the international physics olympiad.

Although the participation in the physics olympiad is highly regarded by the Minister of Education, good results in the contest are neither requested nor helpfull for the continuing studies after highschool.

EXTRA TRAINING OF THE NATIONAL TEAM.

To give an extra training in laboratory work, an other week at
one of the organising universities is organised during which the
students will do experimental tests of former (inter)national physics
olympiads and some lectures about error estimation and experimental
set-ups are given.

In the beginning of the dutch physics olympiad the organisation
didn't provide such a training. By the years it was considered
necessary to give an extra training in order to give the members of
the dutch team a fair chance to compete with team members of the other
countries participating in the international physics olympiad.
Although laboratory work at highschool in the Netherlands is becoming
more and more important this part of teaching is still underscoring
with respect to the teaching of theory. Several reasons are to be
found for this phenomenon amongst which tradition is probably one of
the most important. Other reasons are lack of equipment, difficulties
to raise the necessary funds to buy new equipment and the last but not
the least reason: lack of experience of the teachers to teach physics
by the help of laboratory work. Even the obligation since 1980 that
the final test in highschool schould contain at least one experimental
test didn't change the situation profoundly. So the organisation of
the national physics olympiad found it apt to organise an extra
training especially in experimental work. For the five students
forming the national team this is an extra activity for which they
have to ask a special permission from the staff of their school. So
far this didn't cause any serious problems since all the schools were
quite willing to co-operate, showing by this their interest in
promoting their school as a whole and the physics olympiad in
particular.

Although we are aware of the fact that it is quite impossible to
learn how to do experimental work in a five-days session we are
nevertheless satisfied with the results of it. The selected students
are able to learn very quick and are therefore able to pick up the

most necessary features of laboratory work such as the representation of the mesurements, error estimation, making correct graphs and drawing correct conclusions. Developing a feeling for the experiment and being inventive are out of our training aims.

Dutch Physics Olympiad
participants and schools

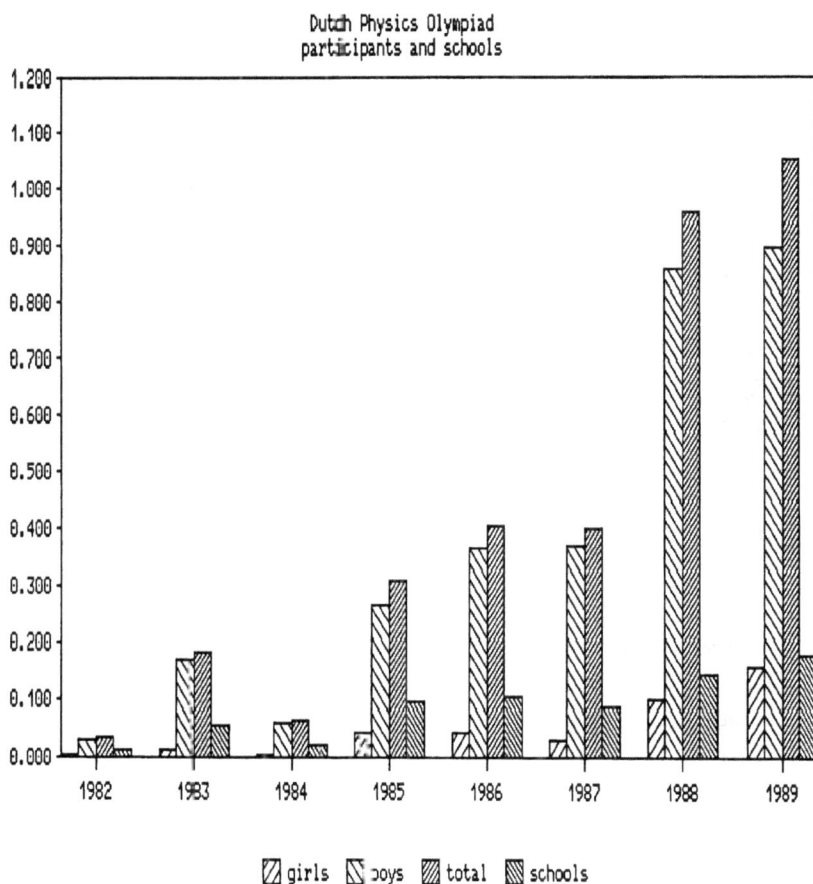

SOME SELECTED THEORETICAL PROBLEMS IN THE DUTCH PHYSICS OLYMPIAD.

1. Some people are able to keep a bottle upside down by just
    pressing their thumb and forefinger to it. Friction plays of
    course an important role.
    Calculate under which conditions this phenomenon is possible.

Solution 1.

In general holds
$$F_w = f.F$$
So
$$F_2 = F_{w1} = f.F_1$$
Since
$$F_g + F_1 = F_{w2}$$
$$= f.F_2 = f^2.F_1$$
we find
$$F_g = (f^2 - 1).F_1$$
This is only possible if
$$f^2 > 1$$
and therefore
$$f > 1$$

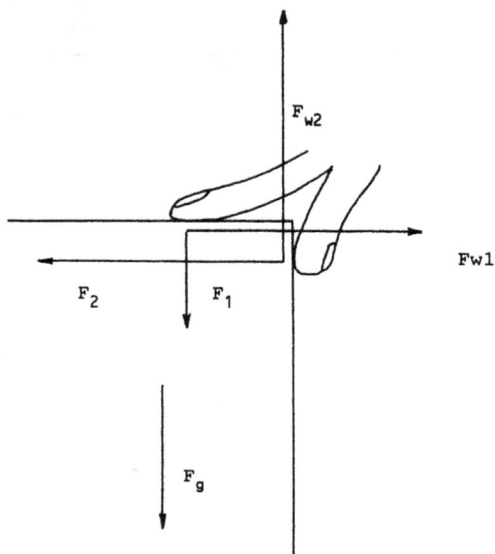

2. Rainbows exist due to refraction and internal reflection in the
    raindrops. Let's consider the case of only one internal
    reflection.
a. Show that the angle between the incoming and the outgoing ray of
    light has a maximum which depends on the wavelength.
b. Explain the existence of the rainbow.

Solution 2.
a. Notice that the paths of the rays of light are symmetrical.
    From the definition of the angles in the drawing it follows

$$\alpha = 2.[r - (i - r)] = 4.r - 2.i$$

and therefore

$$d\alpha/di = 4.(dr/di) - 2$$

From

$$\sin(i) = n.\sin(r)$$

it follows

$$n.\cos(r).(dr/di) = \cos(i)$$

So

$$dr/di = \cos(i)/[n.\cos(r)]$$
$$= [1 - \sin^2(i)]/[n^2 - \sin^2(i)]$$

From

$$d\alpha/di = 0$$

it follows

$$dr/di = \tfrac{1}{2}$$

So

$$4.[1 - \sin(i)^2]) = n^2 - \sin(i)^2$$

and

$$\sin(i) = \surd[(4 - n^2)/3]$$

with

$$n \approx 1,33$$

it follows:

$$i \approx 59,6^\circ \quad ; \quad r \approx 40,4^\circ \quad ; \quad \alpha \approx 42,4^\circ$$

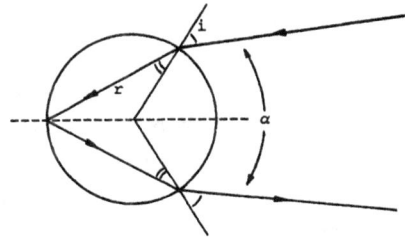

b.   Important elements in the explanation are:

- α_{max}, blew < α_{max}, red
- within the angle of ca 42° colours overlap, so one doesn't see the colours separately; however, at the edge one does.
- light is reflected at all angles within the maximum; however, the intensity of the reflected light has a maximum near the edge.

3.   Just above the horizon the sun seems to be bigger than when it is high in the sky.

Show by calculation whether refraction of light can count for this phenomenon or not.

Solution 3.

Suppose for simplicity the atmosphere to be homogeneous. Being ϕ the real angle under which light from the sun comes in with respect to the vertical, and ϕ' the angle under which the sun is seen from the earth, i the angle of incidence and r the refractive angle then it holds:

$\phi = \phi' + i - r$

since

$\sin(i) = n.\sin(r)$

and $i \approx r$ and $n \approx 1$

we have

$(n - 1).\sin(r) = \sin(i) - \sin(r)$
$= 2.\sin((i - r)/2).\cos((i + r)/2)$
$\approx (i - r).\cos(r)$

so

$(i - r) \approx (n - 1).\tan(r)$

Call the real angle of the sun:

$d\phi$

and the apparent angle:

$d\phi'$

then the enlargement is the ratio

$d\phi'/d\phi$

So

$d\phi'/d\phi = 1 - d(i - r)/d\phi = 1 - (n - 1).d[\tan(r)]/d\phi$

With R = radius of the earth and D = thickness of the atmosphere we have

$\sin(r) = [R/(R + D)].\sin(\phi')$

so we can show that

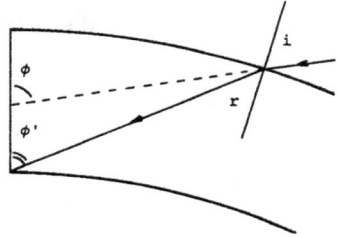

- high in the sky, with r small

    $\tan(r) \approx r \approx [R/(R + D)].\phi'$

  so

    $d\phi'/d\phi \approx 1$

- and near sunset, with

    $\phi' \approx \pi/2$

  it follows that

    $dr/d\phi \approx 0$

  so

    $d\phi'/d\phi \approx 1$

So according to refraction there is no enlargement and one must conclude that therefore the observation is false.

4.   Estimate the minimum speed necessary for a waterskier to be able to ski on one foot only.

Solution 4.

One can show that the force by the water acting on the foot is

    $F = 2.\sigma.A.v^2.\sin^2(\alpha)$

where

  α is the angle between the foot and the water level,

  v is the speed with respect to the water,

  A is the surface of the foot and

  σ is the density of water.

One can also show that F has a maximum for $\alpha \approx 50°$.

Suppose the force of gravity on the skier to be F_g. The forces with respect to the vertical will balance if

    $F_g = F.\cos(\alpha)$

So for a skier with

    $F_g = 800N$ and $A = 0,01m^2$

one finds that the minimum speed is about 10m/s.

5. In Canberra (Australia) exists a fountain that projects water to an altitude of 150 meter. Any moment there is 6 m^3 water in the air.

Calculate the minimum power needed by the pumps.

Solution 5.

Suppose the speed of the water at the bottom of the fountain to be v_0 , then one can write for the power of the pumps

$$P = \frac{1}{2}.\sigma.V/t.v_0^2$$

where

$$V = 3m^3$$

and t is the time needed for the water to go up to the top of the fountain.

Clearly

$$t = v_0/g.$$

From conservation of energy it follows

$$v_0 = \sqrt{[2.g.h]}$$

so

$$P = 3/2.\sigma.g.\sqrt{[2.g.h]} \approx 0,8 \text{ MW}$$

6. Regard the following electrical circuit

$R_1 = 200$ kΩ
$R_2 = 10$ kΩ
$C = 50$ μF
$V_b = 200$ V
N = neon lamp

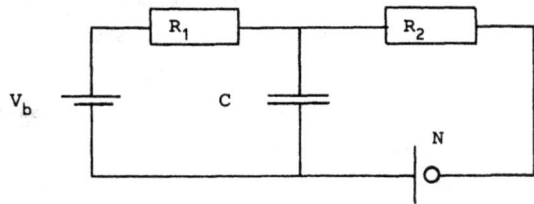

The neon lamp has the following properties:

- it starts to glow at a tension of 100 V
- it extinguishes when the intensity is less than 9 mA.

Calculate the frequency at which the neon lamp blinks.

Solution 6.

To solve the problem we separate both circuits. This is allowed since the RC-time of the two circuits differ a factor 20. The neon lamp starts to burn at a tension of 100V and extinguishes when the current is less than 9mA, i.e. when the tension on the capacity is less than 90V.

Therefore the frequency can be calculated by calculating the time needed to unload the capacity from 100V to 90V over the resistance of 10kΩ and to calculate the time needed to reload it from 90V to 100V over the resistance of 100kΩ.

The unload time

$$t_1 = 0.5*\log(100/90) \approx 0.053s$$

The load time

$$t_2 = 10*\log[(200 - 90)/(200 - 100)] \approx 0.953s$$

So

$$t_1 + t_2 \approx 1.006s$$

and the frequency is therefore $\approx 0.99Hz$.

7. Calculate the equation of the curved part of a flat-curved lens under the condition that a bundle of light parallel to the main axis of the lens is focused at one point.

Solution 7.

Using Fermats principle the optical path is constant for different rays of light. So

$$n.x + \sqrt{[y^2 + (f - x)^2]} = \sqrt{[h^2 + f^2]}$$

for any point (x,y) on the surface of the lens and f being the focal length and 2h the diameter of the lens.

If one uses the approximation

$$h \ll f$$

the expression can be reduced to the form

$$y^2/a - (x - p)^2/b = 1$$

which represents a hyperboloid.

8.   Two identical cylinders are placed parallel in a horizontal
     position. They rotate with the same speed but in opposite
     directions. On top of the cylinders there is a homogeneous bar.
     The friction forces between the bar and the cylinders are
     proportional to the normal forces.
     Obtain the equation of motion of the bar.

Solution 8.

     Suppose the distance between the two cylinders to be L. Then it
     can easily be shown that the net force on the bar, due to
     friction, is proportional to the displacement of the bar out of
     the centre. Therefore the motion is harmonic.
     Since the ratio of the net force and the displacement equals

     $2.f.g/L$

     the angular frequency is

     $\sqrt{[2.f.g/L]}$

     being f the coefficient of friction and g the gravitational
     acceleration. From this the equation is obvious.

9.   Light is refracted from air into a transparent medium. The angle
     of incidence is nearly $\pi/2$. The refractive index of the medium
     depends on the depth in the medium.
     Calculate the dependence of the refractive index with the depth
     as the trajectory of the light is a parabola.

Solution 9.

     Suppose the refractive index n to be a function of y: $n(y)$.
     For any point (x,y) in the medium
     holds

     $n(y).\sin(i) = \text{constant} = n(0)$

     Since

     $\alpha = \tfrac{1}{2}\pi - i$ and $\tan(\alpha) = dy/dx$

     we find

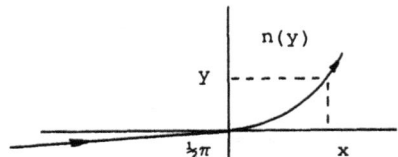

$$n(y) = n(0).\sqrt{[(dy/dx)^2 + 1]}$$

The trajectory of the light is a parabola, so

$$y = a.x^2$$

and therefore

$$n(y) = n(0).\sqrt{[4.a.y + 1]}$$

10. An electrical cirquit contains
   four elements: two capacities,
   one ohmique resister and a ideal
   coil. One applies a sinusoidal
   signal between A and B. The
   outgoing signal is measured
   between A and D.

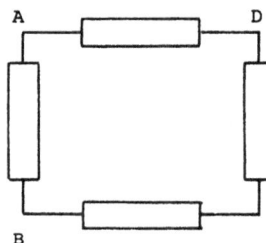

   Determine where the four components have to be placed in order to
   get a filter.

Solution 10.

   The ratio between the ingoing and the outgoing signal equals

   $(Z_1 + Z_2)/(Z_1 + Z_2 + Z_3)$

   From the fact that this ratio has to be zero at a certain
   frequency one can deduct that $(Z_1 + Z_2)$ must be an imaginary
   number. On the other hand $(Z_1 + Z_2 + Z_3)$ must contain a real part
   since the ratio needs to remain finite.

   Therefore Z_1 can be the resister, Z_2 and Z_3 the capacity and the
   coil and Z_4 needs to be a capacity.

11. A tin is filled with water. In the bottom of the tin is a tiny
   hole.
   Calculate the height of the water in the tin as a function of
   time.

Solution 11.

Suppose

the height of the water: $h = h(t)$,

the speed at which the water runs out of the tin: v and

the ratio between the diameters of the tin and the hole in the bottom: p. $(p \ll 1)$

From conservation of energy it follows that

$(dh/dt)^2 + 2.g.h = v^2$

From conservation of mass it follows that

$(dh/dt) = p^2.v$

And since $p \ll 1$ it follows that

$v^2 \approx 2.g.h$

And so

$dh/dt = p^2. \sqrt{[2.g.h]}$

By integrating this equation one finds that

$h(t) = \{\sqrt{h(0)} - p^2. \sqrt{[g/2]}.t\}^2$

12. A massive ball just rests at the edge of a table. There is no friction between the ball and the table. One gently pushes the ball so it rolls off the table.

Calculate the speed of the ball the moment it doesn't touch the table anymore.

Solution 12.

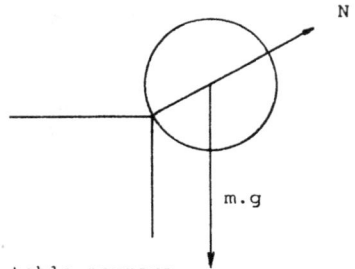

1. the movement of the centre of gravity of the ball is circular, so

$m.v^2/R = m.g.\cos(\alpha) - N$

2. From energy conservation one finds

$\tfrac{1}{2}.m.v^2 = R.[1 - \cos(\alpha)].m.g$

The moment the ball doesn't touch the table anymore

$N = 0$.

By substituting $\cos(\alpha)$ one finds

$v = \sqrt{[2 \cdot g \cdot R/3]}$

13. An one open ended tube is held upside down in a basket filled with water. On the lower side of the tube a weight is connected. When the tube is pushed down in the water there is a position at which the tube remains in equilibrium.
    Determine whether this equilibrium is stable or unstable.

Solution 13.

The problem can be solved without any calculation. The volume of the air trapped in the tube decreases with the depth at which the tube is held in the water.

So the Archimedes force also decreases with the depth. The net force is therefore upwards when the tube is held above the point of equilibrium and downwards when it is held below it. The equilibrium is therefore unstable.

14. Walking on a beach one can observe that the waves are always rolling on parallel with the beach.
    Supposing the speed of water waves to be proportional with the depth of water, and the slope of the beach to be constant, calculate the trajectory of the waves.

Solution 14)

This problem is similar to the
problem in which light passes
through a medium with a variable
refractive index.
There one can write

$n(x) \cdot \sin(\alpha) = $ constant

Since the propagation speed of the waves is proportional to the

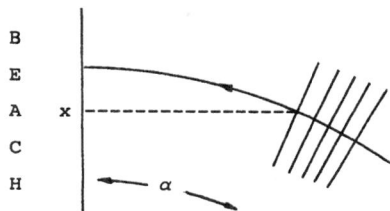

depth and the slope of the beach is constant, $\sin(\alpha)$ is proportional to x, where x is the distance to the beach. So at $x = 0$ we find that $\alpha = 0$ and this explains why the waves roll on parallel to the beach.

Since

$\sin(\alpha) = x/k$ and $\tan(\alpha) = -dy/dx$

it follows that

$-dy/dx = x/[\sqrt{(k^2 - x^2)}]$

Integrating this equation one finds that

$[y - (y(0) - k)]^2 + x^2 = k^2$

This is the equation of a circle with radius k and centre $(0, y(0) - k))$.

15. A nonhomogeneous cylindrical float sticks 10 cm deep, vertically in the water. When one hits the float lightly, it starts to oscillate.

Calculate the period of the oscillation.

Solution 15.

It can easily be proven that the oscillation is harmonic and that for the period holds

$T = 2\pi \cdot \sqrt{[L/g]}$

where

$L = 10$ cm

so that

$T \approx 0{,}6s$

16. Show that the formula for thin lenses holds for a so called Fresnel-zône lens, and that such a lens is chromatic i.e. the focal length depends on the wavelength.

Solution 16.

Suppose an object on the main axis at a distance v from the lens. One receives a clear picture of this object at a distance b from the lens where all the waves arrive in phase. For waves passing the lens at a distance h from the main axis the following equation holds

$$\sqrt{[v^2 + h^2]} + \sqrt{[b^2 + h^2]} - v - b = n.1$$

where l is the wavelength and n is a whole number.

Using the approximation

$$h/v \ll 1 \quad \text{and} \quad h/b \ll 1$$

we find

$$1/v + 1/b = 2.n.1/h^2 = 1/f$$

which is the equatior for thin lenses and clearly the focal length f depends on the wavelength.

17. Parachutes of type 'square' are very much used nowadays because of their good flight properties. They can be regarded as a wing of an airoplane. Turns can be easily made by pulling one of the two steering ropes. The effect is that as well the 'lift' as the air friction are influenced. Air friction is a force contrary to the flight direction and lift is a force perpendicular to that. In general one supposes both forces to be proportional to the speed squared; i.e.

$$F_{lift} = C_{lift} \cdot v^2$$

and

$$F_{friction} = C_{friction} \cdot v^2$$

The constants C_{lift} and $C_{friction}$ depend on how much one pulls the steering ropes. Let's use a parameter r for it, varying between 0 and 1. Suppose the following relations

$$C_{lift} = 6 - 12.(r - 0,5)^2$$

$$C_{friction} = 1 + 1,5.r$$

Calculate the value of r for which a parachutist can fligh the largest distance.

Solution 17.

One can fligh the largest distance when the angle of flight α with the horizontal is the smallest possible.

One can easily show that

$\tan(\alpha) = C_{friction}/F_{lift}$

A minimum is to be found for $r \approx 0,27$.

18. Standing on the ice of a lake I could hear a helicopter approaching. In the beginning the intensity of the sound became stronger and reached a maximum when the helicopter was visible under an angle of about 30°.

After that the intensity of the sound became weaker untill it became stronger again.

Explain the observation and make as good as possible an estimation of the frequency of the sound.

Solution 18.

The phenomenon has clearly to do with interference of the sound reached by the ear directly from the helicopter and sound reflected by the ice.

Suppose the altitude of the helicopter was H and my length is h. Suppose furthermore the horizontal distance to the helicopter to be L. Now the difference in distance travelled by the sound directly to the ear and via the ice can be calculated

$\sqrt{[(H + h)^2 + L^2]} - \sqrt{[(H - h)^2 + L^2]}$

using the approximation

$h/L \ll 1$

this difference can be rewritten as

$2.h/\{\sqrt{[(L/H)^2 + 1]}\}$

Since

$L/H = \sqrt{3}$

the difference is

$2.h/2 = h$

Now because of the reflection on the ice the sound gets a phaseshift of 1/2, and therefore the difference must be half the wavelength to produce an interference maximum. The wavelength of the sound was thus twice my length i.e. 3.8m.

And the frequency of the sound was therefore about 95Hz.

19. Calculate the draw of a cylindrical chimney when there is no wind.

Solution 19.

There is a net force on the colum of air in the chimney due to the fact that there is a temperature difference of the air in the chimney and outside it.

The gravitational force and the Archimedes force are both proportional to the inverse of the absolute temperature.

Let T_0 be the temperature outside the chimney and T_1 inside it. Then it can be shown that the net force is

$$F = C.[1/T_0 - 1/T_1]$$

where

$$C = M.p.L.A/R$$

with M = the molocular weight of air

p = the pressure of the air

L = the length of the chimney

A = the cross section of the chimney

R = the universal gas constant ≈ 8.31 $K^{-1}.mol^{-1}$

When we define the draw of a chimney as the volume of air per second that leaves the chimney then it follows from Poiseuille's law that the draw is proportional to the difference in pressure between both ends of the chimney.

So the draw is proportional to

$$(M.p.L/R).[1/T_0 - 1/T_1]$$

20. Calculate the height at which a billiard ball has to be hit to let it roll without slipping.

Solution 20.

Let the ball have a mass m and a radius R. Suppose one hits the ball horizontally at a height x above the table with a force F during a time t. Does the ball gain a speed v, from concervation of momentum it follows

$$m.v = \int F.dt$$

and

$$F.x = I_A.dw/dt$$

where w is the angular speed.

So

$$m.v.x = \int F.dt.x = \int F.x.dt = I_A. \int dw = I_A.w$$

with

$$I_A = 2.m.R^2/5 + m.R^2 = 7.m.R^2/5$$

and

$$v = w.R \quad \text{(since the ball doesn't slip)}$$

one finds

$$m.w.R.x = 7.m.R^2.w/5$$

so

$$x = 7.R/5$$

SOME SELECTED EXPERIMENTAL PROBLEMS IN THE DUTCH PHYSICS OLYMPIAD.

1. Determination of the refractive index of water.

    By means of a positive lens an object in a basket filled with
    water can be projected scharply on a screen. Suppose one shifts
    the object in the basket away from the screen over a distance x.
    To obtain a sharp picture again one has to shift the basket with
    the object together towards the screen over a distance y.
    Determine from the graph of y versus x the refractive index of
    water.

Solution 1.

    A typical set of measurements is:

x (cm)	0.0	1.0	2.0	3.0	4.0	5.0	6.0	7.0	8.0	9.0
y (cm)	0.0	0.6	1.3	2.2	2.8	3.5	4.3	5.0	5.7	6.4

    From the graph y versus x one can calculate from the slope the
    refractive index
        $n = 1.39 \pm 0.02$

2. Determination of the reduction factor in a rope winded around a
    bar.

    While mooring a ship the moorings are winded a few times around
    the mooring-mast. By doing so one let the moorings slip in order
    to prevent it from breaking from a sudden shock. The force
    necessary to hold the moorings is a fraction of the force acting
    on the ship. The ratio F_1/F_2 between the two forces depend on the
    number of times N the moorings are winded around the mooringmast.
    We call this ratio the reduction factor A.

One can show that

$$F_1/F_2 = A^N$$

In the experiment a rope is winded a few times around a bar. The forces of tension in both ends of the rope are being measured with the help of a couple of weights and a dynamometer.

Solution 2.

A typical set of measurements is:

F_2	F_1	F_1/F_2	F_1	F_1/F_2	F_1	F_1/F_2	F_1	F_1/F_2
		N = 1		N = 2		N = 3		N = 4
60	20	0,33	10	0,16	3	0,05	-	-
110	40	0,36	20	0,18	5	0,05	-	-
160	60	0,38	30	0,19	10	0,06	-	-
210	80	0,38	40	0,19	15	0,07	-	-
270	105	0,39	50	0,19	20	0,07	-	-
320	125	0,39	55	0,17	25	0,08	10	0,03
370	140	0,38	65	0,18	30	0,08	12	0,03

Making a graph of $\log(F_1/F_2)$ versus N one can verify the relation and calculate from the slope the reduction factor

$$A = 0,74 \pm 0,03$$

3. Determination of the temperature by which the Leydenfrost effect takes place in alcohol.

The evaporation of a liquid depends highly of the temperature. The higher the temperature, the faster the liquid evaporates. This is true for a drop of liquid on a hot plate uptill a certain temperature by which a cushion of vapour is formed between the drop and the plate. At that point the 'lifetime' of the drop increases dramatically.

With the help of a heater, a thermometer, oil, ice, alcohol, a
syringe, a thermocouple and a voltmeter, determine the
temperature at which the Leydenfrost effect takes place.

4. Make a filter, consisting of a coil and a capacity, that filters
   at 1500Hz with a bandwidth of 100Hz.
   The bandwidth of such a filter is given by
   $$\delta f = R/(2\pi.L)$$
   in which
   R is the ohmique resistance and
   L is the coefficient of induction of the coil.
   Use the collection of coils and capacities, a generator, a
   voltmeter and a amperemeter.

Solution 4.
   One has to choose the correct coil first, since the bandwidth
   depends only of the coil. This can be done by measuring the
   impedance of different coils with a direct current and an
   alternating current. From the value of L, the value of the
   capacity needed can be calculated. From measuring the impedance
   of different capacities with an alternating current the correct
   capacity can be found and the filter built.

5. A very light rod is fixed at one end so that it can work as a
   pendulum. On the rod a heavy weight is attached in such a way
   that it can be moved up and down. Depending on its position on
   the rod, the period of the pendulum is different. In one position
   the period turns out to be minimum.
   Determine from the minimum of the period the gravitational
   acceleration g.

Solution 5.

Writing down the equation of motion for this physical pendulum one finds for the period

$$T = (2\pi/g).\sqrt{[(a/x) + x]}$$

where a is a constant depending on the mass.

T has a minimum for

$$x_{min} = \sqrt{a}$$

so

$$T_{min} = (2\pi/g).\sqrt{[2.x_{min}]}$$

By measuring T_{min} and x_{min} one can calculate g.

6. Friction between a steel ball and a rail.

In the experimental set-up we use a rail which is slightly bent in the middle. The rail is fixed horizontally so that the lower part is down. A steel ball is released from one side so that is rolls to and fro. Due to friction the movement is slowed down. In the experiment we suppose the friction to be constant.
Determine from the movement the coefficient of friction between the steel ball and the rail.

Solution 6.

The rail has to be fixed in such a way that the angle between both sides of the rail and the horizontal is the same. This can be reached by releasing the ball from both sides and observe where it will stop on the other side. If the positions where the ball stops are not the same the rail has to be adjusted.
If the starting position on one side of the rail is $x0$ and the stopping position on the other side is x_1, it can easily be shown from conservation of energy that

$$F.(x_0 - x_1) = F_w.(x_0 + x_1)$$

and therefore

$$x_1 = [(1 - a)/(1 + a)].x_0$$

with

$$a = F_w/F$$

F being the gravitational force along the slope

$$F = m.g.\sin(\alpha)$$

Every time the ball goes up and down the rail, the new position is reduced by a factor

$$(1 - a)/(1 + a).$$

So by measuring the different positions of the ball on the slope where it stops, on can determine

$$a = F_w/F.$$

On the other hand

$$F_w = f.m.g.\cos(\alpha)$$

so

$$f = a.\tan(\alpha)$$

NORWAY

THE NORWEGIAN PHYSICS COMPETITIONS.

1. ORGANIZATION OF TESTS.

In Norway around half of those who have completed 9 years of compulsory schooling go on to general (academic) secondary school (here called gss) at the age of 16. The first year of gss is common to all pupils, whereas in the two last years several subjects are optional. Physics is among them, and can be studied for one or for two years. Around one fifth of the pupils in gss take the full two years' course.

For the olympiad competition, two tests - both theoretical - are held during the school year. In principle every pupil taking physics in gss can take part, but in practice very few who are not in their second year of physics do so. The local teacher selects the pupils who take part in the first test, and is also given the task of grading this test by the help of a given grading scheme.

The number of schools participating in the first test has been steadily increasing since 1984 when the Norwegian competitions started, but still it comprises only about one fifth of all the schools having a gss curriculum.

Both tests are held at the local schools. The first, in the autumn, takes two and a half hours and consists of many problems ranging from quite simple ones to the more complicated. The purpose is twofold: not to frustrate the pupils, but at the same time to get a good basis for selection. The teachers send the results (those they consider worthwhile) to the olympiad delegation leaders, and we determine the number of pupils to go on to the second test. This number has varied between 50 and 100.

The second test, held in February, takes five hours and consists of somewhat more difficult problems. It also covers a larger syllabus than the first, namely most of the syllabus for the whole two years' course. This test is graded by the delegation leaders, and we pick out the five winners to go on to the international olympiad, with two or three reserves. These seven or eight students get as their prize one year's subscription of Scientific American.

The five who will go abroad are encouraged to study (on their own) the topics from the olympiad syllabus that are not dealt with in their school syllabus. Regrettably we lack a suitable textbook to cover this gap, so they have to find their way in more extensive literature.

In the beginning of June the five go to a week's training course, for the last two years held at Oslo University. The course consists of lectures in some of the topics not covered in the school syllabus, of fairly elementary problem solving in these topics, and of laboratory work.

After that week they are left to themselves until olympiad time.

2. SELECTED PROBLEMS.

a)

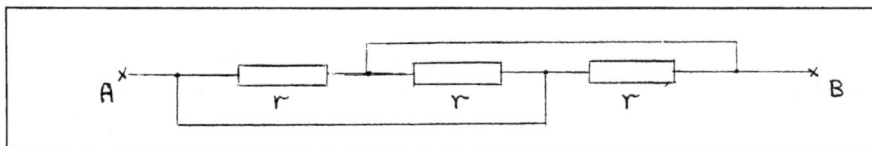

Figure 1

Determine the resistance between the points A and B in fig.1.

Answer: parallel connection gives $R = r/3$.

b)   Two light bulbs, for 220V/40W and 220V/100W respectively, are connected in series to a voltage of 220V. From which bulb will the emission of light be most intense?

Answer: $R_{40}/R_{100} = 2.5$.
In series: $U_{40}/U_{100} = R_{40}/R_{100} = 2.5$

In that case $P_{40}/P_{100} = (U_{40}/U_{100})^2/(R_{40}/R_{100}) = 2.5$, so one would expect the 40W bulb to give more light than the other. Comments about temperature and light emission might give an extra point.

c)

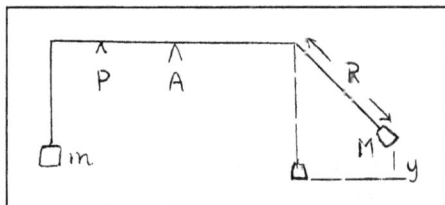

Figure 2

Fig.2 shows a rod that can turn around an axis A at its middle. It is supported at P. Find from what heights y the mass M can be dropped without the rod being lifted from P, when R = 0.50m, m = 5.0kg, and M = 4.0kg.

Answer: $mg \geq S$ and $S - Mg = Mv^2/R$ and $1/2 \cdot Mv^2 = Mgy$ which gives $mg \geq Mg(1 + 2y/R)$ and further $y \leq R/8 = 6.3cm$.

d)   A railway engine has the mass $6.0 \cdot 10^4$kg. What will be the upper limit to the mass of the carriages it can pull on a horisontal track? The coefficients of friction is 0.15 against sliding and 0.0050 against rolling.

Answer: $0.15 \cdot 6.0 \cdot 10^4kg\cdot$g $\geq 0.0050(M + 6.0 \cdot 10^4kg) \cdot$g
$M \leq 1.7 \cdot 10^3$kg.

e)   A monochromatic source emits light in the visual region. When this light falls perpendicularly on a grating with $2.0 \cdot 10^5$ lines per meter, maxima are observed at 22° and at 30° from the central maximum. Determine the wavelength of the light.

Answer: $d \cdot \sin\theta = n \cdot \lambda$ for maxima.
$\sin 22°/\sin 30° = 3/4 = n_1/n_2$ and $400nm < d \cdot \sin\theta/n < 800nm$ leads to
$n_1 > d \cdot \sin 22°/800nm = 2.3$ and $n_2 < d \cdot \sin 30°/400nm = 6.2$ and therefore

$n_1 = 3$, $n_2 = 4$, and $\lambda = 625nm$.

f)

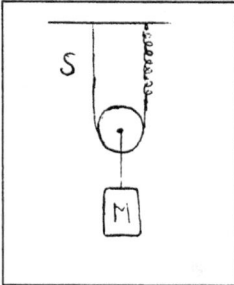

Figure 3

A mass M is supported by a free pulley on the wire S, which is attached to the ceiling at one end and to an elastic spring at the other, see fig.3. The spring will extend 2.0mm per newton stretching force. With M at rest the spring is extended 98mm. The masses of pulley, wire and spring are negligible.

M is slowly pushed upwards until the spring tension is zero, then it is released. How far will M sink before it returns upwards? Find the period of the oscillation that follows.

Answer: Amplitude of spring = 98 mm gives amplitude of M = 49mm, which means that M will sink 98mm from where it was released. The "spring constant" of the system will be $k = Mg/49mm$. Hence the period $T = 2\pi \cdot (M/k)^{1/2} = 2\pi \cdot (49mm/g)^{1/2} = 0.44s$.

g)

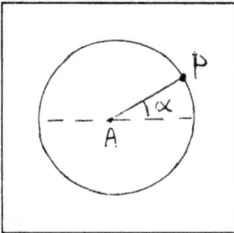

Figure 4

A small, heavy ball is attached to a light, strong string fixed at A, see fig.4, and moves in a vertical circle. The string becomes slack when the ball reaches the point P, where $0° < \alpha < 90°$. Prove that the time needed by the ball to reach its maximum height will be 1/4 of the time the string remains slack.

Answer: Choose A as origo, the x-axis horizontal and the y-axis vertical. v_o = velocity at P. h = maximum height above P. t_h = time from P to maximum height. The ball follows a parabola until it again reaches the circle at Q. t = time from P to Q. r = radius of circle.
The string becomes slack when $v_o^2/r = g \cdot \sin\alpha$.
$v_o^2 = 2gh$ and $t_h = v_o \cdot \cos\alpha/g = (r\sin\alpha/g)^{1/2} \cdot \cos\alpha$

$x = r\cos\alpha - v_o\sin\alpha\cdot t$ and $y = r\sin\alpha + v_o\cos\alpha\cdot t - gt^2/2$

At Q : $x^2 + y^2 = r^2$ and by this, using $\sin^2\alpha + \cos^2\alpha = 1$, we get

$(rg\sin\alpha)^{1/2}g\cos\alpha\cdot t^3 = g^2t^4/4$ and $t = 0$ or

$t = 4\cdot(r\sin\alpha/g)^{1/2}\cdot\cos\alpha = 4t_h$, independent of α.

h)   Parallel light rays in air fall onto the plane side surface of a transparent half sylinder at an angle of 45°. The refractive index is $2^{1/2}$. From where on the sylinder surface will light emerge ?

Answer: $\sin 45° = 2^{1/2}\cdot\sin\alpha$, which gives $\alpha = 30°$, and $2^{1/2}\cdot\sin\beta = 1$, so $\beta = 45°$ and $u = 75°$, see fig.5.

Light rays striking left of P will be totally reflected at least twice and will emerge from SB. Light rays striking right of Q will likewise be reflected and emerge from SA. From the sylinder surface light will emerge in the region PQ , where ASP = 75° and BSQ = 15°.

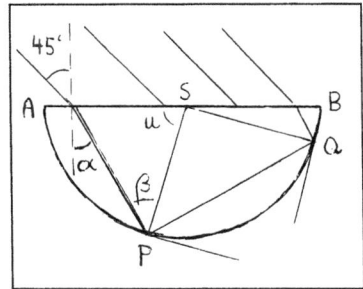

Figure 5

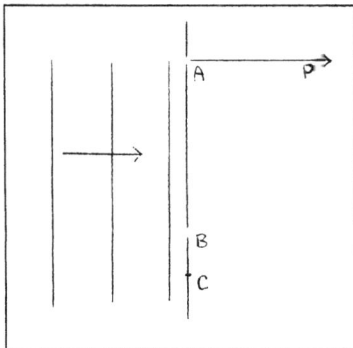

Figure 6

i)

From a distant source sound waves of wavelength λ perpendicularly hit a wall with two small openings A and B at a distance 3λ from each other, see fig.6. A detector is moved along the line AP and thereby encounters two maxima. Determine their distances from A , in terms of λ. Will the intensity at these points increase if a third opening in the wall is made at C, a distance 0.8λ from B (3.8λ

from A) ?

Answer: AP = x . Maximum when
$(x^2 + 9\lambda^2)^{1/2} - x = n\lambda$, where n < 3.
n = 1 gives $AP_1 = 4\lambda$, n = 2 gives $AP_2 = 5/8 \cdot \lambda$.
With opening at C : At P_1 the phase difference = $(CP_1 - AP_1)/\lambda = 1.52$,
which means nearly opposite phases, and reduced intensity. At P_2 the phase
difference = $(3.8^2 + (5/8)^2)^{1/2} - 5/8 = 2.75$. Provided the source gives
sine waves or nearly so, the wave at P_2 can be approximated by $a\sin\omega t +$
$b\sin(\omega t - 3\pi/2) = a\sin\omega t + b\cos\omega t = (a^2 + b^2)^{1/2} \cdot \cos(\omega t - \phi)$, which shows
that the amplitude and the intensity will increase.

j)   A skater increases his speed at a constant rate from 2.0m/s to
12.0m/s along a semicircular turn of length 100m. What is his speed
halfway through the turn? At this point, what will be the angle between
his speed and his acceleration?

Answer:
100m = $1/2 \cdot (2+12)$m/s$\cdot$t   gives t = 100s/7 and a_s =
0.7m/s^2. $v_{50} = (v_o^2 + 2a_s s)^{1/2} = 8.6$m/s.
$\tan\phi = v^2/ra_s$ gives $\phi = 73°$, see fig.7.

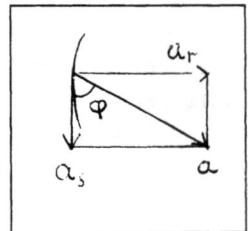

Figure 7

k)

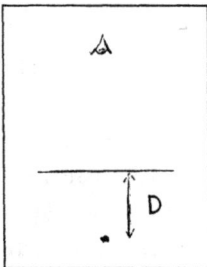

Figure 8

A small coin lies at the depth D in a puddle. Seen
from above (fig.8) it seems to be at the depth d below
the surface. Determine an approximate expression for d.
($\sin\phi \approx \tan\phi$ for small angles.)

Answer: See fig.9. $n \sin\alpha = \sin\beta$, $x/D = \tan\alpha$, $x/d = \tan\beta$, $d/D = \tan\alpha/\tan\beta \approx \sin\alpha/\sin\beta$, and this gives $d \approx D/n$.

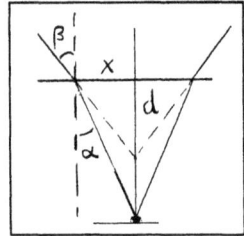

Figure 9

1)

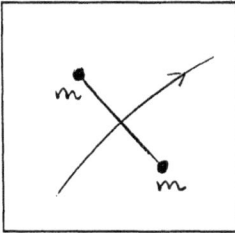

Figure 10

Fig.10 shows an artificial satellite where two equal masses are connected by a strong wire of length 2x and of negligible mass. It follows a circular orbit with radius r = 2R around the earth, where R is the earth's radius. x<<R. Determine the tension in the wire, in terms of m, r, x and g_o. Then let x = 100m, m = 100kg, and calculate the work needed to bring the two parts together.

Answer: Tension = S , earth's mass = M.

Whole satellite: $2m \cdot 4\pi^2 r/T^2 = \gamma 2mM/r^2$

Outer mass: $m4\pi^2(r+x)/T^2 = S + \gamma mM/(r+x)^2$

Eliminating T gives $S = \gamma mM((r+x)^3 - r^3)/r^3(r+x)^2 \approx \gamma mM3x/r^3 = 3mg_o x/4r$.

Work $= 2_0\int^{x=100m} S dx = 0.59J$.

Experimental problems are not given in the Norwegian tests. The closest we ever came is this thought experiment:

You are alone in a room with a blackboard, chalk and a yardstick. Describe a method to find the distance between the pupils of your eyes. Estimate the error in your method. **Answers:** Many!

Ingerid Hiis Helstrup

POLAND

POLISH PHYSICS OLYMPIADS

W. Gorzkowski

Instytut Fizyki PAN, al. Lotników 32/46, PL 02-668 Warszawa

1. GENERAL

The Physics Olympiad in Poland is regularly organized every year since 1951/1952 to develop interest in physics among secondary school pupils, to improve teaching of physics and to help universities in selecting the best candidates for science and technical faculties. Of course, like in many other countries, also earlier there were organized different local competitions on physics or partly on physics, but they did not involve the whole territory of the state and their role was not too great.

The competition is organized in four stages. The introductory and first stages take place in the secondary schools. By the end of September all the schools obtain from the National Board for Physics Olympiad a booklet and posters with both theoretical and experimental problems. The participants solve the problems at home and have a choice of tasks which serves two general purposes. First of all as the students of all the grades obtain the same set of problems the younger ones, who are familiar with only a part of the syllabus, have a better chance to choose the problems which suit them. In addition, the students can develop their interests in various parts of physics chosen by themselves.

The number of participants in the introductory and first stages varies in time from 1000 to 5000. The problems of the introductory stage have a training character and are marked by the school teachers who discuss with the students the problems and their solutions.

The solutions to the problems of the first stage are marked by the members of the thirteen Regional Boards for Physics Olympiad. The authors of the best solutions (about 70% of the participants of the first stage competition) are invited to the theoretical part of the second stage competition organized in thirteen regions at the same time (med of January). The problems are prepared by the National Board for Physics Olympiad and are identical for all the participants (without possibility of choice). The best participants (about 40%) are invited after about a month to solve the experimental problem of the second stage competition.

The solutions are marked by the Regional Boards and later again by the National Board. About 80 students who obtained the best total scores in the second stage (for both theoretical and experimental parts) are invited to the third stage which is organized in Warsaw by the National Board (March/April). The final stage, like all the other stages, involves both theoretical and experimental problems.

The winners of the final stage (about 25 students) get diplomas and prizes. Prizes are also awarded to the teachers of the winners and the schools of the winners obtain special olympic diplomas.

The participants of the third stage can enter the physics faculties of all the universities in Poland without any entrance examinations. The winners can enter the physics and mathematics faculties of the universities and all the faculties of the technical universities without any examinations. These privileges vary slightly from year to year and on occasion they have even been larger.

The marking schemes for the first, second and third stages are prepared by the National Board and each solution is marked by at least two persons independently. This is justified by the above mentioned privileges and by strong competition among candidates when entering universities.

Each of the Regional Boards consists of about 10 persons. But only one of them (Secretary) gets a salary. The National Board consists of about 20 people. The salary is paid to: Director, Scientific Secretary (recently there are two Scientific Secretaries: one for theoretical problems and one for experimental problems), Secretary, Accountant and Office-boy.

Both the National Board for Physics Olympiad and the Regional Boards for Physics Olympiad are usually attached to the universities and consist mainly of the university teachers, scientists from the Polish Academy of Sciences, secondary school teachers and representatives of the educational administration. It is worthwhile to mention here that, although the leading role is played by science, the cooperation between science, schools and educational administration is very close. Without this cooperation the Physics Olympiad could not work efficiently.

The Physics Olympiad has its budget consisting practically only of the financial support from the Ministry for National Education. We should, however, underline that all the Boards (the Regional and the National ones) make use of the lecture rooms, laboratories and offices of their home universities free of charge.

One of the most important points in the work of the National Board are very detailed discussions on various problems (for the current and next Olympiads), their solutions, marking schemes, etc., organized practically every week from September/October until February/March. The sessions of the National Boards are like seminars.

In principle, everybody can be the author of an olympic

problem as the National Board buys the problems to the Bank of Problems independently of the status of their authors. In fact, among the authors there are academicians, university professors, teachers, students, and even pupils. However, most problems are composed by the members of the National Board, in particular, by the Scientific Secretary(ies). We should emphasize that all the problems are refereed by people with rich olympic experience.

The training system of the Polish team changed in time. At present we invite the ten best winners to the Warsaw University for about 10 days. The training involves both theory and experiment, but the experimental part is much more important as the experimental equipment in many home schools of the winners is insufficient. At the end of the training the participants pass some tests and then the final decision concerning participation of the candidates in the olympic team is taken.

The purpose of the training is to improve the experimental skill of the students and to help them in solving the problems which are in the International Physics Olympiad Syllabus and which are missing from the Polish secondary school one. During the training the problems of the past International Physics Olympiads as well as the problems of the past (Polish) Physics Olympiads (published every 2 - 3 years in a book form) are used.

The final selection at the end of the training provides additional motivation for more efficient work before and during the training.

Some typical theoretical problems of the Polish Physics Olympiads are presented below.

2. THEORETICAL PROBLEMS

Problem 1

A glass tube with internal and external radii equal r and R, respectively, is filled with a luminescent liquid which, under the influence of X-rays, emits a green light. For the green light the refractive indexes of the glass and of the liquid are n_1 and n_2, respectively.

What condition must be fulfilled by the ratio r/R so that one seemed from the outside as if the thickness of the glass wall were equal to zero?

Solution

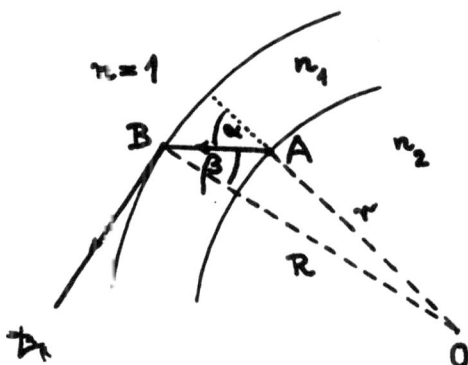

Fig. 1

The impression that the thickness of the glass wall is zero arises if and only if the external rays, coming from the liquid, reach eye, running along a tangent to the outside surface of the glass.

For cylindrical symmetry of the system it is sufficient to consider only one ray leaving the system along a tangent to the tube and perpendicular to the axis of the tube (Fig. 1).

The rays which pass from the liquid on to the point A are bent into the dihedral angle $2\alpha_{max}$. For the ray running along the tangent to the tube the incident angle β is the critical angle for the air-glass surface. Thus,

$$\sin \beta = \frac{1}{n_1}.$$

The sine theorem applied to the triangle ABO gives the relation:

$$\frac{r}{\sin \beta} = \frac{R}{\sin(180° - \alpha)},$$

i.e.

$$\frac{r}{R} = \frac{\sin \beta}{\sin \alpha}.$$

The required condition is met when

$$\alpha \leq \alpha_{max},$$

i.e. when

$$\frac{r}{R} \geq \frac{1/n_1}{\sin \alpha_{max}}.$$

The value of α_{max} depends on the ratio of the refractive indexes n_1 and n_2. We have:

a) if $n_2 \leq n_1$, then

$$\sin \alpha_{max} = \frac{n_2}{n_1}$$

and then

$$\frac{r}{R} \geq \frac{1}{n_2}.$$

b) if $n_2 \geq n_1$, then

$$\sin \alpha_{max} = 1$$

and then

$$\frac{r}{R} \geq \frac{1}{n_1}.$$

The effect similar to the one described can be seen while looking at a bottle filled with milk. In spite of the fact that the bottle glass is thick it seems to us that the milk files the bottle from the "left" external surface to the "right" external surface.

Problem 2

Before two touching soap bubbles merge, there is often an intermediate stage with a thin film between the bubbles (Fig. 1).

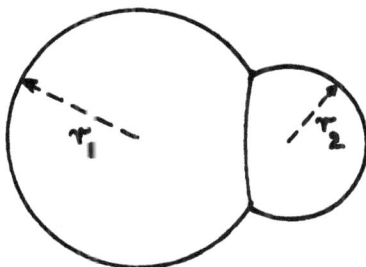

Fig. 1

a) Find the radius of curvature r_{12} of the film that separates the bubbles if the radii of curvature r_1 and r_2, shown in Fig. 1, are known.

b) Consider the special case of $r_1 = r_2 = r$. What were the radii of the bubbles before the intermediate state was formed? What is the radius of the bubble after the intermediate film vanished?

We assume that the excess pressure in the bubble depends only on the surface tension and the radius, and is much less than the atmospheric pressure, so that the total volume of gas in the bubbles does not change.

The volume of a spherical dome (Fig. 2) is:

$$\frac{1}{3} \pi \ (2r^3 - 3r^2 d + d^3).$$

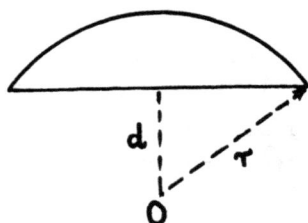

Fig. 2

Solution

The only expression, with dimension of pressure, which can be formed of the surface tension σ and the radius of curvature r is:

$$A \frac{\sigma}{r},$$

where A is a dimensionless constant. Thus, the excess pressure under the soap film with the surface tension σ and the radius of curvature r is:

$$\Delta p = A \frac{\sigma}{r}.$$

From equilibrium of pressures in the intermediate state it follows that

$$A \frac{\sigma}{r_1} + A \frac{\sigma}{r_{12}} = A \frac{\sigma}{r_2}.$$

Therefore,

$$\frac{1}{r_{12}} = \frac{1}{r_2} - \frac{1}{r_1}.$$

Positive values of r_{12} correspond to the situation illustrated in the Fig. 1. In the case of negative r_{12} the intermediate film is convex in opposite direction (to the bubble with the radius r_2).

The surface tension can be treated as a force (per unit of length) which acts on the perimeter along a tangent to the surface of the liquid. Three forces of equal magnitude (the same σ for all the films) are in equilibrium if and only if their directions form angles of 120° with each other. It follows that the planes, tangent to the films, in every "triple" point form equal angles with each other — Fig. 3.

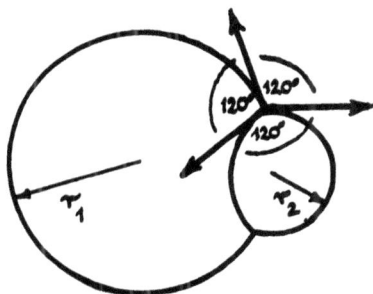

Fig. 3

In a particular case where $r_1 = r_2 = r$ it follows that the distance between the centre of curvature of the bubbles must be equal to r (Fig. 4).

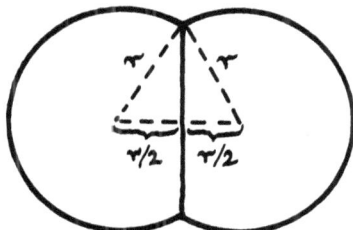

Fig. 4

The dome with $d = r/2$ has volume $V = \frac{5}{24} \pi r^3$. According to the assumption mentioned in the text of the problem the volume of the gas is conserved. Thus,

$$\frac{4}{3} \pi R^3 = \frac{4}{3} \pi r^3 + \frac{4}{3} \pi r^3 - 2 \cdot \frac{5}{24} \pi r^3,$$

where R is the radius of the bubble after bursting the intermediate film. This formula gives:

$$R = \frac{3}{2} \frac{r}{\sqrt[3]{2}}.$$

For radii ϱ of the bubbles before the formation of the intermediate film we obtain:

$$\frac{4}{3} \pi \varrho^3 = \frac{4}{3} \pi r^3 - \frac{5}{24} \pi r^3,$$

$$\varrho = \frac{3}{2} \frac{r}{\sqrt[3]{4}}.$$

It is interesting to note that the formula for Δp was obtained by employing the dimensional analysis only, without difficult calculations, and that the value of A was not important for solving the problem. Another interesting point is that for small values of Δp in the bubbles the total volume of the gas is conserved.

Problem 3

You have two equal portions of water, e.g. 1 l each, at temperatures $0^{\circ}C$ and $100^{\circ}C$, respectively. You have also various vessels with arbitrary volumes and with various properties of their walls (adiabatic and/or diathermic).

You want to warm up the water initially cold with the heat of the water initially warm.

Find the maximum obtainable temperature of the water initially cold. Describe and justify the method you suggested.

Remarks: The two portions of water cannot be mixed with each other. For the sake of simplicity, we assume that the specific heat of water is constant and we neglect the thermal expansion of water.

Solution

The answer is surprising: the final temperature of the water initially cold can practically be equal to $100^{\circ}C$!

At the beginning we describe how the cold water can reach $80^{\circ}C$.

At first we construct a "heat exchanger". For this purpose we take 10 vessels of equal capacity whose total volume is negligible compared to 1 l. We fill 5 of them with the cold water and the remaining 5 with the warm water. Now we establish thermal contacts between the small vessels and the original portions of water (cold or warm) in order to obtain the distribution of temperatures shown in Fig. 1.

No. of the vessel	water initially cold	water initially warm	No. of the vessel	
5	100°			
4	80°	80°	$1'$	
3	60°	60°	$2'$	
2	40°	40°	$3'$	
1	20°	20°	$4'$	
		0°	$5'$	Fig. 1

430

As the total volume of the vessels is negligible, the heat loss when constructing the heat exchanger is negligible too. Thus, the temperatures of the initial portions of water are not changed. They are 0°C and 100°C, respectively. For the same reason, it can be assumed that also the volumes of the initial portions of the water are unchanged.

Now we pour the water with $t = 20°C$ out of the vessel 1 to a larger vessel. It is the coldest water obtained from the water initially warm in the process we are considering. We do the same with the water with t = 80°C: we pour this water out of the vessel 1' into another large vessel. It is the warmest water obtained from the water initially cold in the process considered.

After these operations are completed the vessels 1 and 1' are empty. We fill them with the warm and cold water, respectively, and change the configuration of the vessels into that shown in Fig. 2.

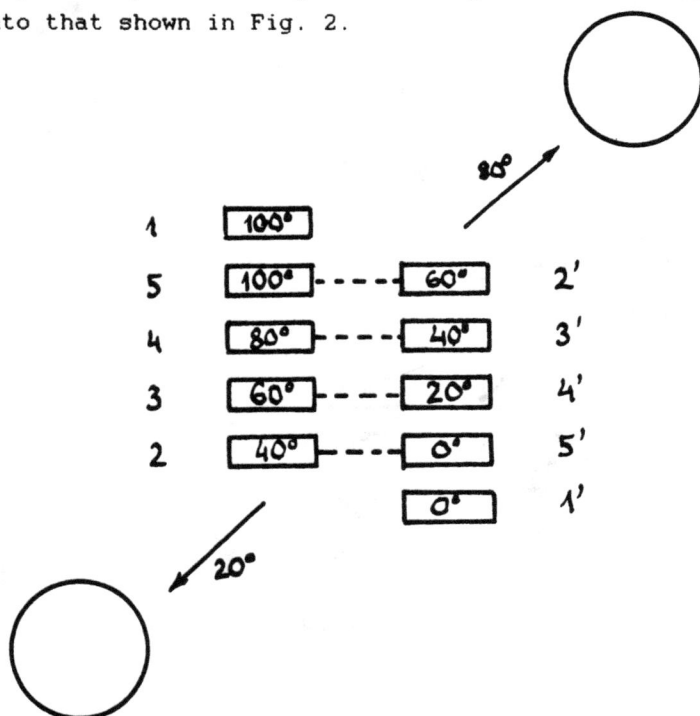

Fig. 2

Now we establish thermal contacts between the vessels "on the same level", marked with the line ------- in the Fig. 2. As a result, we obtain distribution of temperatures which is exactly the same as that shown in the Fig. 1. But, the larger vessels contain one small portion of water initially warm with the temperature 20°C and one small portion of water initially cold with the temperature 80°C.

The process described above can be repeated many times. After each cycle we receive one additional, small portion of water initially warm with $t = 20^{\circ}$C and one additional, small portion of water initially cold with $t = 80^{\circ}$. At the end we receive 1 l of water initially warm with $t = 20^{\circ}$C and 1 l of water initially cold with $t = 80^{\circ}$C.

An important point in our considerations is that after each cycle the state of the heat exchanger (Fig. 1) is reconstructed, i.e. after each cycle the temperature distribution in unchanged.

It is clear that by taking 100 vessels with the total volume $\ll 1$ l we can heat the cold water up to 98°C. For 1000 vessels with the total volume $\ll 1$ l the final temperature of the initially cold water would be 99.8°C, etc.

The problem considered above illustrates how the real heat exchanger works. The only difference is that the real heat exchanger, schematically shown in the Fig. 3, works continuously, while our heat exchanger works in a discrete manner.

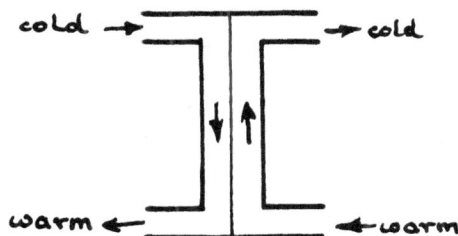

Fig. 3

Problem 4

A homogeneous ball with radius $r = 1$ mm made of an ideally black material is illuminated with a strong, homogeneous, large enough, parallel laser beam (Fig. 1). The laser radiation is circularly polarized and its wavelength is equal to $\lambda = 10$ μm. The ball, initially at rest in the laboratory reference system, starts to translate and rotate. Each point of the ball (except those on the rotation axis) is moving along a helical trajectory. Find the slip H of this helical trajectory. Neglect gravitation (you can assume that the experiment is performed in a space laboratory).

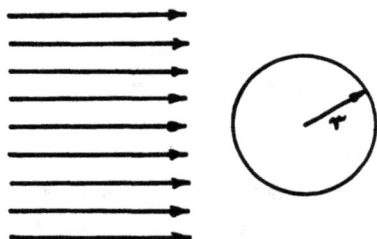

Fig. 1

Remark: The monochromatic, circularly polarized radiation can be considered as a set of moving photons, each with the same energy, the same momentum and the same angular momentum. For each photon the directions of the angular momentum (vector!) and momentum (also a vector) are parallel. The value of the angular momentum of each photon is $\hbar$ ($= h/2\pi$; h – Planck constant).

Solution

During a short time Δt the ball absorbs $n\Delta t$ photons, where n denotes the number of photons falling on the ball in the unit of time. The momentum of each photon is h/λ. In effect, the momentum of the ball increases in the time Δt by

$$\delta p = \frac{h}{\lambda} n\Delta t .$$

But,

$$\Delta p = m\Delta v ,$$

where m denotes mass of the ball and v is its velocity. Hence, the linear acceleration of the ball is:

$$a = \frac{\Delta v}{\Delta t} = \frac{nh}{m\lambda} .$$

Each photon has the angular momentum $h/2\pi$. The absorbtion of the photons by the ball changes its angular momentum by

$$\Delta L = \frac{h}{2\pi} n\Delta t .$$

But,

$$\Delta L = I \Delta \omega ,$$

where L is the moment of inertia of the ball and ω is its angular velocity. It is well known that

$$I = \frac{2}{5} mr^2 .$$

Hence, the angular acceleration ε of the ball is:

$$\varepsilon = \frac{\Delta \omega}{\Delta t} = \frac{nh}{2\pi I} .$$

The ratio of a/ε is:

$$\gamma = \frac{4}{5}\, \pi r^2/\lambda.$$

This quantity is constant in time. This means that the trajectory of each point of the ball is helical independently of the shape of the laser pulse, i.e. independently of how the intensity of the laser radiation depends on the time. Thus, for simplicity's sake, we can assume that the intensity of the laser beam is constant in time. Then not only γ is constant, but both α and ε are constant too, and we have:

$$x = \frac{nh}{2\lambda m}\, t^2,$$

$$\phi = \frac{nh}{4\pi I}\, t^2,$$

where x and ϕ denote the linear and angular positions of the ball, respectively. We have here made use of the fact that for $t = 0$ the ball was at rest.

Therefore,

$$x = \frac{4\pi r^2}{5\lambda}\, \phi.$$

H is equal to the increase of x corresponding to the increase of ϕ by 2π. Thus,

$$h = \frac{8\pi^2 r^2}{5\lambda} \qquad (1.6\ \text{m})$$

Problem 5

26 thin metallic plates with radii r and 26 thin metallic plates with radii R $(R > r)$ are situated parallel

to each other as in Fig. 1.

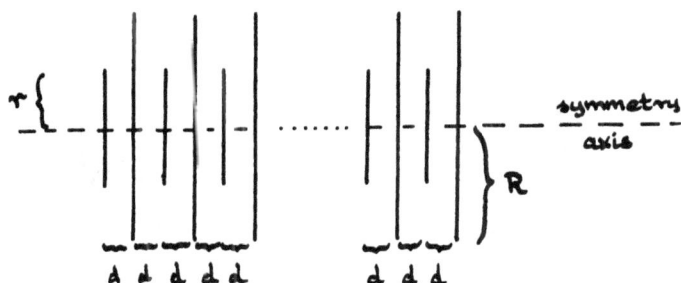

Fig. 1

The distances between any two subsequent plates are equal to d $(d \ll r)$. How should the plates be connected into two sets of plates for the capacity of the capacitor obtained in this way to be the highest? Find value of this highest capacity.

Solution

We should connect some plates to the potential V_1 and the remaining ones to the potential V_2 so that the energy of the electrostatic field, accumulated in the system, were the greatest. Of course, the final result for the capacity cannot depend on the values V_1 and V_2. So, for simplicity's sake, we can put

$$V_1 = V/2, \qquad V_2 = -V/2.$$

We find the energy of the electrostatic field in two cases illustrated in Fig. 2 and 3.

1) (Fig. 2) The surfaces of the plates are S_1 and S_2, respectively. The electrostatic field in the region shown in the Figure is:

$$E_1 = V/d.$$

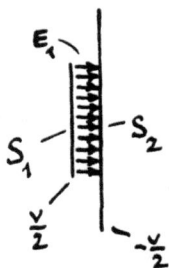

Fig. 2

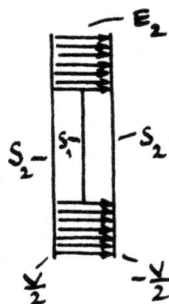

Fig. 3

The energy of the electrostatic field in the region considered is:

$$W_1 = \varepsilon_0 \frac{1}{2} (S_1 d) \left[\frac{V}{d} \right]^2 = \frac{1}{2} \varepsilon_0 \frac{S_1}{d} V^2.$$

2) (Fig. 3) Similarly, as $E_2 = V/2d$, the energy W_2 of the electrostatic field in the region shown in the Figure (without the central part) is:

$$W_2 = \frac{1}{2} \varepsilon_0 (S_2 - S_1) 2d \frac{V^2}{4d^2} = \varepsilon_0 \frac{1}{4} (S_2 - S_1) \frac{V^2}{d}.$$

We remark that $W_1 > W_2$ if and only if

$$S_1 > \frac{1}{3} S_2.$$

Now we consider the energy of the electrostatic fields between two subsequent greater plates. Theoretically, we have 8 possible ways of connecting the plates to the potentials $+V/2$ and $-V/2$. However, only four of them are essentially different. The remaining four connections correspond to the change of sign of both potentials: $+V/2 \leftrightarrow -V/2$.

The possible four cases are illustrated in Fig. 4.

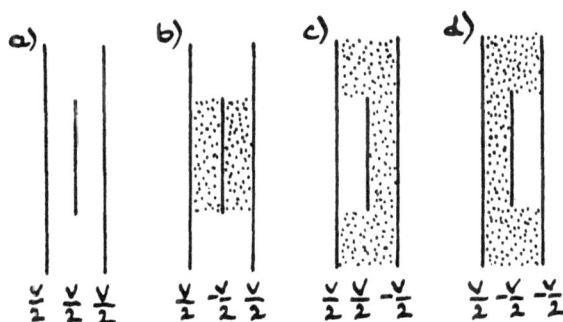

Fig. 4

The energy of the electrostatic field equals to:

case a: 0

case b: $2W_1$

case c: $W_1 + W_2$

case d: $W_1 + W_2$

For $S_1 > \frac{1}{3} S_2$ this energy is the greatest in the case b. For $S_1 < \frac{1}{3} S_2$ it is the greatest in the cases c and d. We see that the "best" connection of the plates should depend on the ratio S_1 and S_2.

For each fixed ratio S_1/S_2 one should find a connection corresponding to the maximum energy of the electrostatic field in all the regions between the greater plates. We have:

A) For $S_1 > \frac{1}{3} S_2$, i.e. for $R < \sqrt{3}\, r$, we connect all the smaller plates together and all the greater plates together – Fig. 5.

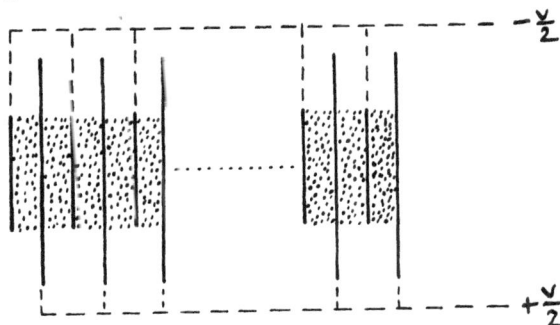

Fig. 5

The total energy of the electrostatic field in this case is:

$$W_A = 51 \, W_1 = \frac{1}{2} \frac{51 \, \varepsilon_o S_1}{d} V^2 = \frac{1}{2} C_A V^2,$$

where C_A denotes the capacity of the system. Thus,

$$C_A = \frac{51 \, \varepsilon_o S_1}{d}.$$

B) For $S_1 < \frac{1}{3} S_2$, i.e. for $R > \sqrt{3} \, r$, the greater plates should be connected to $+V/2$, $-V/2$, $+V/2$, $-V/2$, $+V/2$, $-V/2$, ... (alternately to $+V/2$ and $-V/2$). As regards smaller plates between the greater ones, they can be connected all to $+V/2$, all to $-V/2$ or part of them to $+V/2$ and remaining part to $-V/2$. This case is illustrated in the Fig. 6.

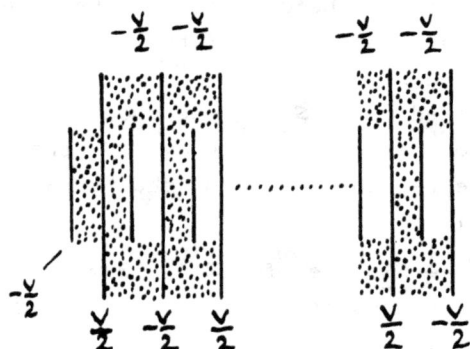

Fig. 6

In this case the energy of the electromagnetic field equals to

$$W_B = W_1 + 25(W_1 + W_2) = \frac{1}{2} \frac{25S_2 + 27S_1}{2d} \varepsilon_o V^2 = \frac{1}{2} C_B V^2.$$

Thus, the capacity in this case is equal to

$$C_B = \frac{25S_2 + 27S_1}{2d} \varepsilon_o.$$

Taking into account that $S_1 = \pi r^2$ and $S_2 = \pi R^2$, we can write:

$$C = \begin{cases} \dfrac{51\pi\varepsilon_o r^2}{d} & \text{for } R \leq \sqrt{3}\, r \\[2ex] \pi\varepsilon_o \dfrac{25R^2 + 27r^2}{2d} & \text{for } R \geq \sqrt{3}\, r \end{cases}$$

Problem 6

An accelerator produces protons with kinetic energy $E = 2$ keV each. A narrow beam of such protons is directed onto a metal sphere with radius r. The sphere is in a large distance from the accelerator. The distance between the centre of the sphere and the initial trajectory of the proton beam is $d = \frac{1}{2} r$ (Fig. 1).

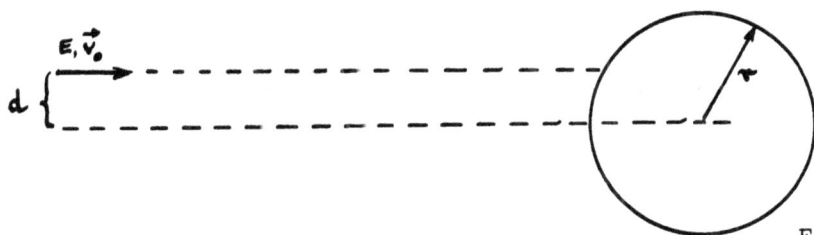

Fig. 1

Assuming that the accelerator is kept running for sufficient length of time, calculate the final potential of the sphere.

We assume that the intensity of the beam is low so that the mutual interaction between the protons in the proton beam can be neglected.

Solution

The protons colliding with the metal sphere transfer their charge to the sphere — they remain in it. In consequence, the sphere becomes charged and its charge is increasing. The interaction between the proton beam and the charged sphere modifies the trajectory of the beam. Finally, the protons move along the path 3 shown in Fig. 2. When this situation is reached then the process of charging the sphere stops. We have to find the electrostatic potential of the sphere (with respect to the potential in infinity) in this case.

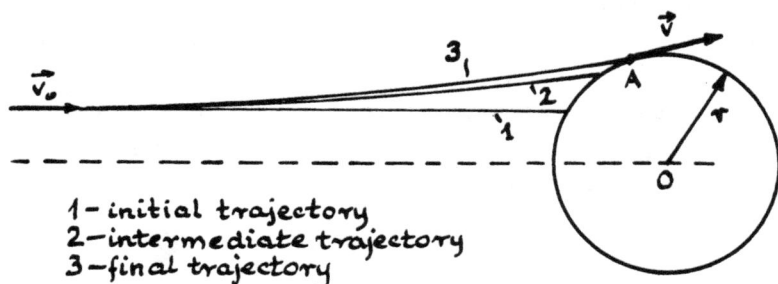

1 — initial trajectory
2 — intermediate trajectory
3 — final trajectory

Fig. 2

Applying the energy conservation law to each proton in the beam we can write

$$E = eU + \frac{1}{2} m\upsilon^2 ,$$

where: m — mass of the proton, e — its electric charge, υ — velocity of the proton in the point A shown in the Fig. 2.

eU is the potential energy of the proton in the point A, and $\frac{1}{2} m\upsilon^2$ is its kinetic energy in that point.

Remark that the force acting on the protons is a central one. The interaction between each proton and the charged sphere is such as if the total charge of the sphere were localized in its geometric centre O. We have assumed here that the charge of the sphere is large enough so that one proton approaching the sphere practically does not modify the charge distribution on the sphere. This assumption is justified by small charge of the proton compared to macroscopic charge of the sphere in the final state and by the small intensity of the proton beam.

As the interaction between the protons and the sphere is central, we can use the angular momentum conservation law:

$$m \, v_0 d = m \, v \, r,$$

where v_0 denotes the initial velocity of the protons, i.e. the velocity of the protons leaving the accelerator. The last formula can be written in the following form:

$$\frac{1}{2} m \, v^2 = \left(\frac{d}{r} \right)^2 E,$$

where $E = \frac{1}{2} m v_0^2$. We use non-relativistic formula for E as E is much less than the proton rest energy $(=mc^2)$.

Combining this formula with the formula describing the conservation of energy law we get:

$$U = \left(1 - \frac{d^2}{r^2} \right) \frac{E}{e} \qquad (= 1500 \text{ V}).$$

It is proper to remark that the protons bombarding the sphere can cause a secondary emission of electrons (but not protons as the protons are bound rather strongly in the crystal lattice). This phenomenon accelerates the process of charging the sphere.

The situation would be quite different if we considered an electron beam instead of the proton beam. In such case the phenomenon described above would discharge the sphere and in the final state the beam would move along some

trajectory like the trajectory 2 in the Fig. 2. The
trajectory 3 would never be reached.

Problem 7

A thin homogeneous inextensible line of length l and
mass M is initially fastened by both ends to two hooks which
are close to each another, and hangs freely as in Fig. 1a.

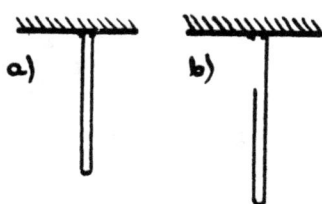 

Fig. 1   Fig. 2

Then one end of the line is released and begins to fall
as in Fig. 1b. The largest load N which each of the hooks
can bear is greater than the weight of the line. What
conditions must mg and N fulfill in order that the second
hook is not ripped out?

We assume that during the fall, as each element of the
line reaches its final position, it remain there motionless.

Solution

First method

In situation shown in the Fig. 2 the velocity of the
left part of the line is $\sqrt{2gh}$. The momentum of the element
AB of the line with length dx is:

$$\frac{M}{l} \, dx \, \sqrt{2gh} \, .$$

This momentum vanishes during a time dt which is equal to the time necessary for passing the way AC. This time is equal to

$$2 \, dx/v \, .$$

i.e. to

$$\frac{2 \, dx}{\sqrt{2gh}} \, .$$

Thus, the additional force acting on the hook. which holds the line, is:

$$\frac{dp}{dt} = \frac{M}{l} \, dx \, \sqrt{2gh} \, \frac{1}{2dx} \, \sqrt{2gh} = M \, g \, \frac{h}{l} \, .$$

For the greatest possible value of h $(h = l)$ this force has its maximum equal to Mg. The maximum value of force acting on the hook is equal to the sum of the weight of the line and the maximum value of the additional force. Therefore, the condition we look for is:

$$N \geq 2 \, M \, g \, .$$

Second method

We shell use the formula:

$$\frac{d\vec{p}}{dt} = \vec{F} \, ,$$

where $\vec{p}$ is the total momentum of the system and $\vec{F}$ is the total force acting on the system (line). The momentum $\vec{p}$ of the line has only vertical component. Its value is equal to the momentum of the left part of the line with the length $(l-h)/2$; we have:

$$p = \frac{l - h}{2} \frac{M}{l} \sqrt{2gh},$$

but

$$h = \frac{1}{2} gt^2,$$

hence,

$$p = \frac{l - \frac{1}{2} gt^2}{2} \frac{M}{l} gt.$$

The total force $\vec{F}$ acting on the line is equal to the sum of the weight of the line $M\vec{g}$ and the force $\vec{R}$ (reaction) shown in the Fig. 2. The force $\vec{F}$ is vertical like the momentum $\vec{p}$. Thus,

$$\frac{d}{dt} \left[\frac{l - \frac{1}{2} gt^2}{2} \frac{M}{l} gt \right] = Mg - R,$$

i.e.

$$\frac{1}{2} Mg - \frac{3}{4} \frac{M}{l} g^2 t^2 = Mg - R.$$

This relation should be fulfilled in each moment when the line is moving. For the time

$$t = \sqrt{2l/g}$$

(which corresponds to the lowest position of the free end of the line), we have:

$$\frac{1}{2} M g - \frac{3}{2} \frac{M}{l} g l = M g - R.$$

Therefore,

$$R = 2 M g.$$

Of course, this is the maximum value of the force R. Thus,

$$N \geq 2 M g.$$

Problem 8

Two ideal gases, monoatomic and biatomic, are mixed with one another and form an ideal gas again. The equation of the adiabatic process of the mixture is:

$$p V^{\varkappa} = \text{const},$$

where $\varkappa = 11/7$.

Let n_1 and n_2 denote the numbers of moles of the monoatomic and biatomic gases in the mixture, respectively. Find the ratio n_1/n_2.

Solution

From the first principle of thermodynamics we have:

$$\Delta Q = \Delta U + p \Delta V.$$

The molar specific heat at constant pressure can be determined by considering one mole of an ideal gas:

$$c_p = \frac{\Delta Q}{\Delta T}\bigg|_{p=\text{const}} = \frac{\Delta U}{\Delta T} + p\,\frac{\Delta V}{\Delta T},$$

where $U = C_v T$ is the internal energy of one mole of the ideal gas (C_v is the molar specific heat at constant volume). We have:

$$\Delta U = C_v\,\Delta T.$$

Using the equation of state (for one mole):

$$p\,V = R\,T,$$

for p = const, we obtain:

$$p\,\Delta V = R\,\Delta T.$$

Thus, from the first equation we get:

$$c_p = c_v + R,$$

where R is the gaseous constant.

It follows from the kinetic theory of gases that the internal energy of ideal mixture of two (or more) ideal gases is equal the sum of their internal energies:

$$U = U_1 + U_2 = n_1\,C_{v1}T + n_2\,C_{v2}T.$$

For the mixture of gases the following relation holds:

$$(n_1 + n_2)\,C_v = (\Delta Q/\Delta T)\big|_{v=\text{const}} = \Delta U/\Delta T = n_1 C_{v1} + n_2 C_{v2}.$$

Therefore,

$$C_v = (n_1 C_{v1} + n_2 C_{v2})/(n_1 + n_2).$$

It is well known that:

$$C_{v1} = \frac{3}{2} R \qquad \text{(for monoatomic gas)}$$

$$C_{v2} = \frac{5}{2} R \qquad \text{(for biatomic gas)}$$

Thus,

$$\varkappa = C_p/C_v = \frac{[n_1 C_{v1} + n_2 C_{v2} + (n_1 + n_2)R]/(n_1 + n_2)}{(n_1 C_{v1} + n_2 C_{v2})/(n_1 + n_2)} =$$

$$= \frac{5(n_1/n_2) + 7}{3(n_1/n_2) + 5}.$$

But we know that:

$$\varkappa = 11/7.$$

Hence

$$\frac{5(n_1/n_2) + 7}{3(n_1/n_2) + 5} = \frac{11}{7}$$

Solving this equation, we obtain:

$$\frac{n_1}{n_2} = 3.$$

SWEDEN

SWEDEN

Lars Gislén ,Department of Theoretical Physics

Sölvegatan 14 A, 223 62 Lund, Sweden

Bosse Lindgren, Department of Physics,

Vanadisvägen 9, 113 46 Stockholm,Sweden

Selecting The Team

Sweden has since 1975 an annual National Physics Competition. This competition is organized by the Swedish Physical Society and a newspaper, Svenska Dagbladet, and is financed by an anonymous donation. Thus the competition is in principle an off-school activity.

The first part of this competition consists of a written examination with eight theoretical problems in late January covering the whole country and with around 1000 participants from about 150 schools. The six best school teams, a team being the three best students of a school, get prize sums. The fifteen best individuals are gathered for a final held in Stockholm in early May. The final consists of one day of three or four experimental problems and the next day a written theoretical examination with eight problems. Each problem can give at most 3 points and the experimental total points are then multiplied by a suitable factor to give theory and experiment equal weight. In case two participants get the same total points the result of the experimental problems will be used to decide which one is the best. All participants in the final are awarded money prizes from 6000 SEK to 750 SEK, but have otherwise no advantage, e.g. when they later apply for entrance to university.

The national competitions are not formally associated with the nomination of an olympic team but the best students in the final will in practice constitute the Swedish team.

Training

The last three years we have trained the team before the IPhO. The main reasons for doing this is that the IPhO curriculum contains topics not included in the Swedish school curriculum such as angular momentum, moment of inertia and adiabatic processes. Another reason is to teach experimental techniques: handling of errors of measurement and how to write a report.The last but not least reason is to get the members of the team together, tell them what happens at an IPhO and to make them relaxed and confident.

The actual training takes place during three days just before the departure for the IPhO. About a month before we have sent the members of the team some twenty pages of theory together with solved examples and also problems for the students to solve for themselves. During the three days we use about half the time to revise this theory, highlighting difficult points and discussing topics that the students bring up. As a rule the afternoons are used for experimental work. We ask the students to work as a group, solving some experimental problems, mostly taken from previous IPhOs. They have to produce a report of their results which we discuss and criticise afterwards. We try to provoke them to make mistakes as we think that this is a good way of learning. The evenings are free for social activities.

The cost ($ 300) of the training is covered by the Swedish Physical Society. The students are lodged in guest researcher rooms at the Lund University which we are able to get at a very low cost during the summer. The university teachers involved (often delegation leaders) give their lectures and tutoring free of charge. The main part of the total cost of training is for food.

Theoretical Problems

1. A cylindrical plastic rod is placed on a lined paper and photographed. The picture shows the result.
a) Explain qualitatively the distortion of the lines.
b) Estimate the refractive index of the plastic by making measurements in the photo.

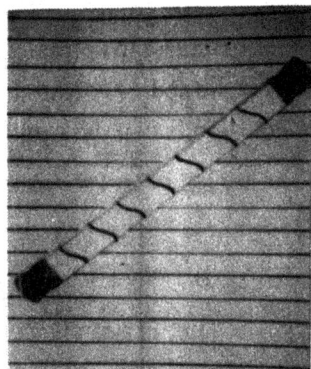

2. At radioactive decays of the uranium series, the decays normally produce radiation with small energy and thus small power of penetration. However, in 0.7% of the decays of U-238 go via an isomer of U-234, producing a high energy photon. This radiation is very penetrating and was studied at the radiation measurements made on the grounded Soviet submarine U137. In the measurements the detector used had an efficiency of 0.25% (i.e. one photon of 400 gives a puls in the detector). The sensitive area of the detector was 22 cm^2 and you may suppose that it was places 1.5 m away from the radiation source. The source can be treated as pointlike. In the measurements 125 pulses/h were registered. Estimate the amount of uranium in the submarine assuming that 93% of the radiation was absorbed in the armour plate of the ship. The half-life of U-238 is $4.5 \cdot 10^9$ years.

3. The spectrum of a star similar to the sun is observed to have a superimposed spectrum from a hotter start. The spectral lines of the two stars are periodically displaced relative to each other. From this one draws the conclusion that we are looking at a double star, the two component of which move around the common mass center.
The hydrogen H$_\alpha$ line (λ=656.2785 nm in the laboratory), emitted from both stars, is studied. When the separation of the lines is maximal the respective wavelengths are 656.3879 and 656.5382 nm. About 20 days later the intensity of the sunlike spectrum decreases suddenly and at the same time the total

light intensity from the double star decreases to a minimum (see figure). Later the light returns to normal intensity. During the minimum the sunlike spectrum disappears and you observe a single H_α line with wavelength 656.4535 nm. The conclusion is that the sunlike star was eclipsed by the other component.

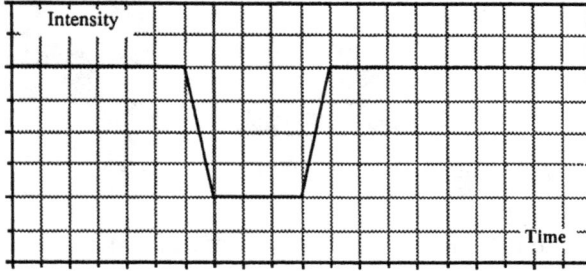

a) Compute the velocity component of the two double star components in the direction to the earth in the different cases.

b) Compute the ratio of the component masses.

c) Compute the ratio of the radii.

Assume that the plane of the orbit is parallel to the line of sight and that the stars move in circular orbits.

(Final 1981)

4. The human heart pumps about 80 ml of blood in each beat. The excess pressure (blood pressure) during the beat is in average 16 kPa (120 mm Hg). Estimate the net power of the heart.

(Qualification 1986)

5. Investigate the following crude model of a nucleus. The nucleus is assumed to be a cube of n x n x n nucleons. Every nucleon is attracted by nuclear forces (strong interaction) from the other nucleons. As the range of this force is very small we may assume that only the nearest heighbours interact with a given nucleon. Each nucleon nucleon pair interaction is asssumed to contribute with a certain constant energy to the total energy of the nucleus.

The nucleus has an electric charge Z. This gives a repulsion between the nuclei.

From dimensional arguments the total electrostatic energy of the nucleus is proportional to Z^2/d where d is the linear extension of the nucleus. In this crude model we may assume that Z is proportional to A, the number of nucleons in the nucleus.

The suggested model gives a relation between the binding energy per nucleon and the number n. Derive this relation.

It is known that nuclei in the neighbourhood of iron (Fe) in the periodic table are very stable and have binding energies of about 8.7 MeV/nucleon. Use this fact to draw a diagram of binding energy per nucleon as a function of A.

(Final 1986)

6. Mountaineers sometimes use a technique of climbing between two vertical rock walls that is illustrated in the figure. Assume that the coefficient of friction between the shoes and the rock wall is 1.2 and that the length of the leg of the person is 0.9 m.

a) Determine the distances between the walls for which the mountaineer can stand in the position shown.
b) Give some suggestions of how to cope with other distances.
(Qualification 1987)

7. One morning last December, as I was jogging on the outskirts of Stockholm, my path led me past the wall of a large factory. Following an impulse, I clapped my hands and listened. Istead of the expected echoing handclap I heard what at first seemed to be a sparrow emitting a swiftly descending chirp. After several more claps and chips I realized that it was no sparrow – besides, where was the echo?

For the next few minutes I stood in front of the factory clapping and listening. For every clap there was a chirp. I realized with astonishment that is was as the handclap (a superposition of sound waves with different frequencies) had been sorted according to frequency, the high fequency components echoed back sooner than the low frequency. But how? Finally I noticed that the factory wall was not flat but was made of corrugated metal, the evenly spaced vertically oriented corrugations having a spacing between neighbouring corrugations of about 0.1 m. Therein lay the explanation...

The paragraphs above were taken (slightly changed) from a note by Frank S. Crawford Jr. to the America Journal of Physics 1970, page 378.

a) Explain the phenomenon qualitatively.

b) Make a graph of the time delay versus frequency for the reflected handclap. Assume that you stand 20 meters from the wall.

(Final 1987)

8. A capillary glass tube (length 0.600 mm, inner diameter 2.0 mm) contains a 50 mm mercury column. The column divides the capillary in two parts: vacuum and a mixture of air and water vapour. By tilting the tube, the length of the gas compartment can be varied. An experiment was done with different tilts and the table shows the result.

d/mm	h/mm
521	100
259	200
100	400
68	500
51	600

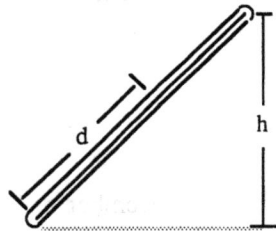

After each measurement the gas was allowed to establish equilibrium. Determin the amount of air and water repectively in the tube.

(Final 1988)

9 Recently contact lenses have been made by utilising the wave character of light. These lenses have alternating dark and transparent concentric rings on their surface around a circular transparent area. See figure. For simplicity we consider the lens as a very thin plane plate with concentric dark and transparent rings.

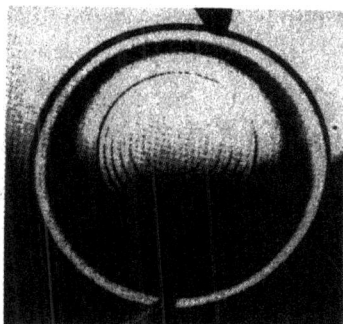

a) Compute the average diameter of the two innermost transparent rings around the central area in such a lens if you want the focal distance to be 250 mm. The lens will be a positive lens.

This lens turns out to me a multifocus lens, being capable not only of focusing distant objects 250 mm from the lens but also object placed at nearer.

b) At what distance from the lens will this be?
((Qualification 1989)

Assume that the light has wavelength 500 nm and that the concentric rings are very narrow.

10. People such as lift repairers, towing specialists and yachtsmen often want to know the size of the tension force in a rope or wire where there is no access to a free end. A British firm now manufactures a meter which clips on to the rope. A lever is operated to introduce a small deflection in the rope (See figure). The meter then straighforwardly measures the restoring force orthogonally to the rope at M. Derive an expression from which the tension of the rope can be calculated. Calculate the tension if the deflection is 12 mm and the restoring force 300 N. (Qualification 1986)

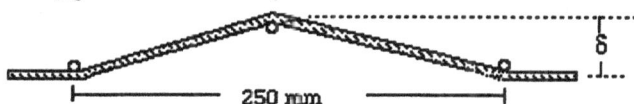

250 mm

Solutions Of The Theoretical Problems

1.a) Consider the image of a point A in the figure. The image will not be displaced in the direction of the axis but will fall somewhere on CD. Perpendicularly to the axis the rod behaves as a magnifying glass and the image of A will be farther away from the axis in point B. The image of FE will appear at GH.

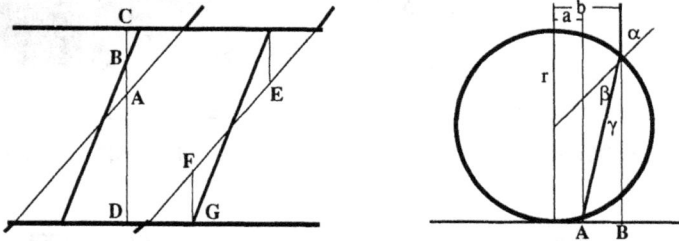

b) If A is close to the axis (central rays) all angles are small and we have

$$\sin\beta \approx \beta = \frac{b}{r} , \quad \sin(\beta - \alpha) \approx \beta - \alpha = \frac{b-a}{2r}$$

$$\Rightarrow \alpha = \frac{b+a}{2r}$$

The refraction law gives

$$n = \frac{\sin\beta}{\sin\alpha} \approx \frac{\beta}{\alpha} = \frac{2b}{b+a}$$

If we draw a straight line through GH and also a perpendicular to the cylinder axis as shown in the figure below, we can measure OA' and OB' which are proportional to a and b. We find OA'=11.2 mm, OB'=23.4 mm which gives $n=1.35$.

If the rays are not central the angles will not be small. This distorts GH from a straight line as seen on the photograph.
(Qualification 1981)

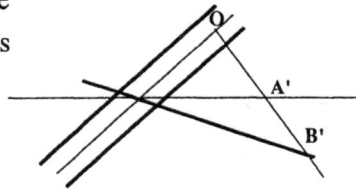

2. The solid angle of the detector is $\omega = (0.0022)/(1.5)^2$.

The total number of disintegrations per hour in the spherical angle is

$$\frac{dN}{dt} = \frac{125 \cdot 400 \cdot 4\pi(1.5)^2}{(1 - 0.93) \cdot 0.07 \cdot 0.0022} h^{-1} = 1.15 \cdot 10^{16} / \text{year}$$

We also have

$$\frac{dN}{dt} = -\lambda \cdot N \text{ with } \lambda = \frac{\ln 2}{T}$$

which gives $N = 7.46 \cdot 10^{25}$ uranium atoms.

The amount of uranium is $\dfrac{7.45 \cdot 10^{25} \cdot 238}{6.023 \cdot 10^{26}} \text{ kg} = 29.5 \text{ kg}$

The amount of uranium in the submarine was about 30 kg.

3. The changes in wavelength is due to the movements of the star relative to the observer (Doppler effect). We have $v = c \, \Delta\lambda / \lambda$.

a) During the eclipse we measure the velocity of the mass center. The formula above gives $v = 80$km/s.

At maximal separation the stars move relative to the mass centers with velocities $v_1 = 50$ km/s and $v_2 = 119$ km/s.

b) The momentum is conserved which gives $(m_1 + m_2) \cdot V_{cm} = m_1 v_1 + m_2 v_2$. This gives $m_1/m_2 = 1.3$. The mass of the hot star is 1.3 times the mass of the sunlike star.

c) A study of the eclipse intensity curve gives (see figure) $r_1/r_2 = 1/4$.

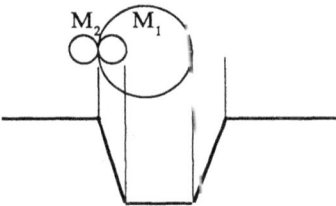

4. Assume a heart beat rate of 70 beats/minute. Assume the area of the aorta is A and that the blood is displaced the distance d during the beat. The force on the blood is $p{\cdot}A$ and the work done is $p{\cdot}A{\cdot}d = p{\cdot}V$.

The work of one beat is then 16 kPa $\cdot$ $8{\cdot}10^{-5}$ m^3=1.28 J.

The power is $1.28{\cdot}70/60$ W=1.5W.

With a reasonable value of the area of the aorta the kinetic energy of the blood of the order of 0.01 J which can be neglected here.

5. By some thought you find that there are $3n^2(n-1)$ pairs of strong bindings. The electrostatic energy is proportional to $Z^2/d \sim n^5$. The total binding energy per nucleon is then $3a(1-1/n)-b{\cdot}n^2$, where a and b are constants to be determined. From the text we get

$$3a(1-1/n)-b{\cdot}n^2 = 8.7 \qquad \text{(Maximim binding energy)}$$
$$3a/n^2-2bn = 0, \text{ with } n=(56)^{1/3} \qquad \text{(Maximum for Fe)}$$

which gives

$$b = 0.131 \text{ MeV and } a = 4.798 \text{ MeV}.$$

You get the binding energy diagram which compares favourably with reality in spite of the crude model.

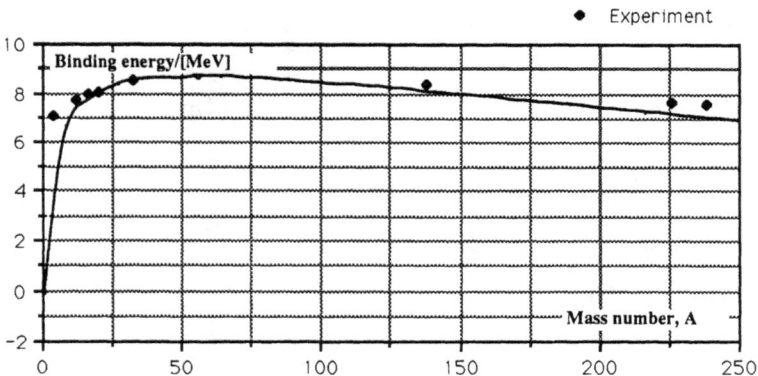

6.

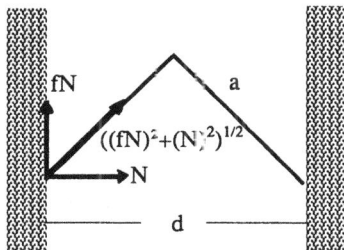

Denote leg length with a and wall distance with d.

a) For the moutaineer to 'hang' between the walls the reaction force from the respective wall must be directed along her legs. The geometry of the figure gives (f being the coefficient of friction):

$$N / ((fN)^2 + N^2)^{1/2} = d / (2a) \text{ or}$$
$$d = 2a / \sqrt{(1 + f^2)}.$$

The largest possible value of f is 1.2 that is $d \geq 1.2$ m.

We also naturally have $d \leq 1.8$ m.

b) The figures show some alternative positions.

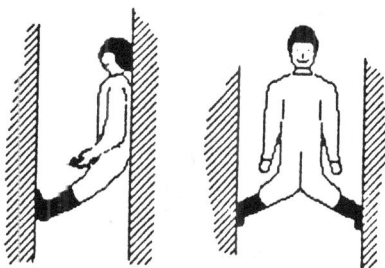

7.

a) The corrugated wall makes a diffraction grating reflecting lower frequencies at larger angles. The low frequency waves will then have a longer distance to travel.

b) Denote the grating spacing by d. From the figure we have that the reflected sound waves are in phase when $2\,d\sin\alpha = 1$. The factor 2 comes from the wave going forth and back.

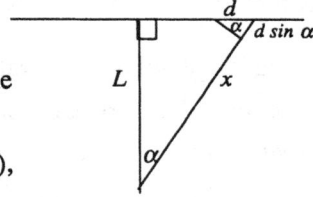

Consider first order diffraction only. The distance is $2x = 2\,L/\cos\alpha$..

The time to cover this distance is $2L/(c\cos\alpha)$, where c is the speed of sound.

With frequency $f = c/\lambda$ we get

$$\Delta t = 2L \cdot c \frac{2fd}{\sqrt{4f^2d^2 - c^2}}$$

We get a lower cut off frequency $f_0 = c/(2d) = 1.7$ kHz corresponding to reflection from very far away (small amplitude). When frequency increases t decreases toward $2L/c$ when $f \rightarrow \infty$.

The phenomenon is particularly strong at the frequencies for which the ear is most sensitive.

Higher diffraction orders give frequencies one octave higher with the same time delay. These will be hard to separate.

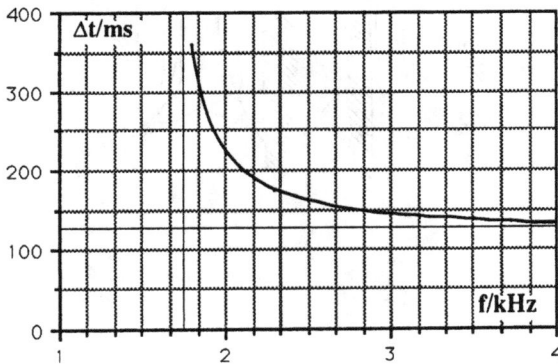

8. An ideal gas obeys Boyle's law, at constant temperaure $p \sim 1/V$. A suitable way to represent the data is to draw p versus $1/V$. We then find that the data splits into two parts, one part being a straight line through the origin. For higher pressures the data lie on a straight line not passing the origin. As long as the water vapor in the tube is not saturated, the gas will behave like a perfect gas. When the water starts to condense, the water vapor partial pressure is constant while the air partial pressure follows Boyle's law. This explains the data.

For $p < p_{saturation}$ we have

$$p \cdot V = (n_A + n_W) \cdot R \cdot T \text{ or } p = (n_A + n_W) \cdot R \cdot T \cdot 1/V$$

where n is the number of moles of air (A) and water (W) respectively.

When $p > p_{saturation}$ we have

$$p = n_A \cdot R \cdot T \cdot 1/V + p_{saturation}.$$

The problem text does not give the temperature T. One way is to assume that the room temperature is 20 C. An error in this temperature of 10 gives only a 3% error in the absolute temperature. Another and better way is to exptrapolate the saturation pressure from the data and then get the temperature from a standard tabl. From the two lines one can then compute $n_A + n_W$ and n_A respectively. You get $2.9 \cdot 10^{-7}$ mol air and $4.4 \cdot 10^{-7}$ mol water or 8.6 g air and 7.9 g water.

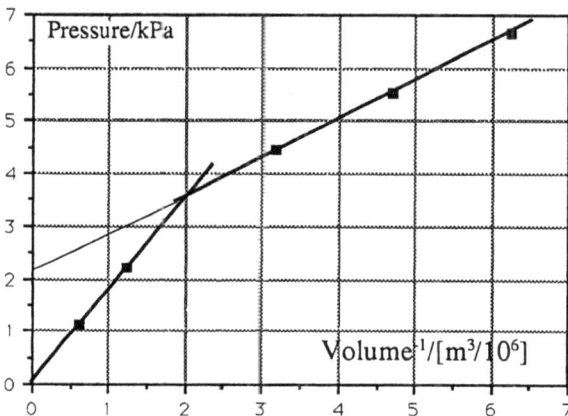

9. To have constructive interference at the point P at distance f from the lens, the path difference between a ray passing the lens center and a ray passing the ring with radius r_n must be a multiple of the wavelength.

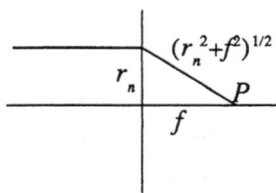

$$\sqrt{r_n^2 + f^2} - f = n\lambda$$

This gives

$$r_n^2 + f^2 = (n\lambda + f)^2$$

$$r_n = \sqrt{(n\lambda)^2 + 2\lambda nf} \approx \sqrt{2nf\lambda} \quad \text{as} \quad n\lambda << f$$

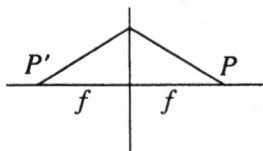

For $n=1,2$ we get $r_1 = 0,50$ mm and $r_2 = 0,70$ mm.

It is easy to see that from a point P', symmetrically situated on the other side of the lens the path difference will be exactly even multiples of the wavelength. A point 250 mm from the lens will also be focused at P.

10. Assume that the tension of the rope is T. Equilibrium of the forces gives

$$F = 2T\sin\alpha \approx 2T\tan\alpha = 2T\delta / a$$

$$T = \frac{Fa}{2\delta} = \frac{300 \cdot 0,125}{2 \cdot 0.012} = 1.6 \text{ kN}$$

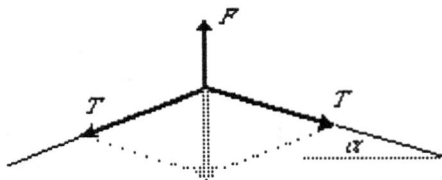

EXPERIMENTAL PROBLEMS

1. (1/1988)   A mousetrap has a strong spring, which is put under tension even in the closed position. To load the trap a certain amount of energy is needed. Find this amount of energy by doing necessary measurements.

Equipment: Mousetrap, string, pulley, weights, balance, standrod, double clamps, clamps to secure the mousetrap to the bench, ruler and protractor.

Solution:
The required energy is equal to the work W to fully load the trap:

$$W = \int F dx$$

If the base of the trap is mounted vertically and the string is attached to the outer part of the moving wire a suitable weight on the string will open the trap a certain angle θ between 0 and 180°. The distance the weight has moved is then $x = l(1-\cos\theta)$ (where l is the lever arm). If the forces (mg) are plotted for a number of distances x (from 0 to $2l$) the work required can be obtained from numerical integration in the diagram. Note that each load usually gives two different angles θ.

2. (5/1981)   A few grams of dry ice is placed inside a toy balloon and the balloon is sealed off. The balloon is placed on a balance. The dry ice will sublime and inflate the balloon. This causes a decrease in the reading of the balance. Make necessary measurements to determine the density of air and gaseous CO_2. Use five balloons to get a sufficient mean value. Try to estimate the accuracy in this method of measurement.

Equipment: Balance, toy balloons, dry ice, graduated glass cylinder, water.

Solution:
The apparent weight of the balloon will decrease due to the action of the surrounding air on the inflated balloon according to Archimedes' principle. The balloon has to be weighed before it has been filled with dry ice (m_b), after it has been filled but before it has been sealed off (m_b+m_{di}), and after it has been inflated m_{app}. The volume of the balloon V can be determined by measuring the volume of water it displaces upon immersion or, if the balloon is sufficient spherical by measuring its diameter.

$$m_b + m_{di} - V\rho_{air} = m_{app} \qquad V\rho_{CO_2} = m_{di}$$

3. (1/1978) A multimeter consists of a galvanometer with an inner resistance R_g connected to two chains of resistors in series according to the figure. The upper chain has the total resistance 7.50 kohm. The first resistor in the lower chain has the resistance 1.75 kohm. Determine the internal resistance of the multimeter for the 2.5 V range. Determine the resistances R_g and R_x.

Equipment: The multimeter, voltage supply, two potentiometers 1-10 000 ohm and 1-100 000 ohm, cables.

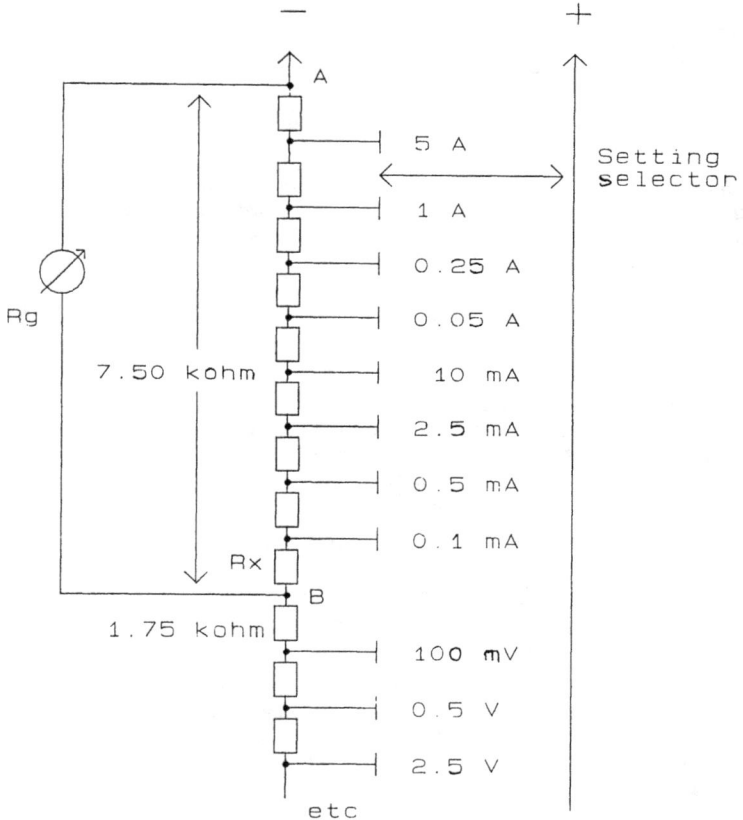

Solution:
a. Measurement of the internal resistance R_i of the multimeter at the 2.5 V range: Connect the multimeter in series with the potentiometer R of 100 kohm to the voltage supply. Start with the potentiometer at 0 ohm. Adjust the voltage so that the meter reads full deflection 2.50 V. Increase R until the meter makes half the deflection 1.25 V. Then $R = R_i = 62$ kohm.

b. Use the same method as above to determine the inner resistance R_1 at the range 100 mV. The measurement will give R_1 = 2.51 kohm. The resistance between the points A and B is R_{AB} = 2.51 - 1.75 = 0.76 kohm.

$$1/R_{AB} = 1/0.76 = 1/R_g + 1/7.5 \Rightarrow R_g = 0.85 \text{ kohm}$$

The voltage over R_g at full deflection 100 mV is $U_g = (0.76/2.51) \cdot 100 = 30.3$ mV. The current through R_g at full deflection is $I_g = (30.3/0.85) = 35.6$ μA.

At the range 0.1 mA the voltage drop over $R_g + R_x$ is the same as over $7.5 - R_x$. At full deflection one gets

$$35.6(R_g - R_x) = (100 - 35.6)(7.5 - R_x)$$

with R_g = 0.85 kohm. This gives

$$R_x = 4.53 \text{ kohm}$$

4. (2/1983 and 2/1989) Five different resistors are connected inside a black box according to this diagram:

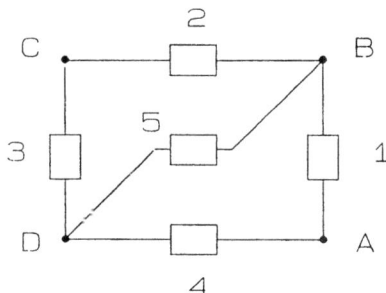

You can only reach the circuit from the four terminals A-D. You are not allowed to open the box. Determine the resistance of resistor #5 (the diagonal). [It is possible to obtain all five resistances.]

Hint: Use the unit kohm. Make all calculations in the quantity $y = 1/R$ in the unit $(\text{kohm})^{-1}$.

Equipment: Multimeter, cables and the black box.

Solution:
Shortcircut some of the points A, B, C and D and measure all resistors in parallel. Use the admittance $y = 1/R$ in the calculations:

Measurement	Shortcircuit	Measure over	y
I	BD	1	$y_1 + y_4$
II	BD	2	$y_2 + y_3$
III	AD,BC	1	$y_1 + y_3 + y_5$
IV	AB,CD	2	$y_2 + y_4 + y_5$
V	AB,BC	3	$y_3 + y_4 + y_5$

$$III + IV - I - II = 2y_5$$

$$\left. \begin{array}{l} III - IV = y_1 - y_4 \\ I = y_1 + y_4 \end{array} \right\} \quad \text{gives } y_1 \text{ and } y_4$$

$$IV - y_4 - y_5 = y_2$$

$$V - y_4 - y_5 = y_3$$

5. (3/1987) A vessel filled with water is rotating on a turntable. The water surface has then paraboloidal shape and at sufficient low rotating speed it is possible to use a spherical approximation at least near the axis of rotation. Determine the gravitational constant (g), by using optical methods.

Equipment: Turntable with vessel, lamp, screen, tape measure and stopwatch.

<u>Solution:</u>
When the lamp is placed a distance R (R is the radius of curvature of the rotating surface) above the surface an image of the lamp can be seen close to the lamp. A water particle of mass m on the surface at a distance r from the axis of rotation will feel a centrifugal force $mr\omega^2$ and a gravitational force mg. When the surface is stationary the resultant of these two forces is normal to the surface.

$$r = R \sin\theta, \quad \tan\theta = \omega^2 r/g, \quad g = \omega^2 r/\tan\theta \approx \omega^2 r/\sin\theta = \omega^2 R$$

If R is determined as above and the angular velocity by using a stopwatch, g can easily be obtained.

6. (3/1979) Build a Michelson interferometer from the given components. Illuminate the interferometer properly with laser light to obtain an interference pattern. Place a 10 mm thick evacuated cell in one arm of the interferometer and let air slowly into the cell to determine the index of refraction for air. The laser light has the wavelength 632.8 nm.

Equipment: Evacuable cell, vacuum pump, laser with beam expander, mirrors, beamsplitter.

Solution:
The difference in optical distance when the cell is evacuated and when it is filled with air is $2(n-1)d$ where n is the refractive index of air and d the thickness of the cell. The difference in optical distance corresponds to a number of wavelengths p. The number p can be determined by continously observing the interference pattern when air is gently let into the evacuated cell.

$$p_{obs}\lambda = 2(n-1)d$$

(p_{obs} = 9 and d = 10 mm gives (n-1) = 0.00028 and n_{air} = 1.00028)

7. (1/1987) Try to determine with the highest possible accuracy the gravitational constant g by doing measurements on the air track. The amount of air is controlled with a variable transformer, preventing the gliders from touching the track. Too much air results in retarding the gliders. A body attached to a spring oscillates with a period

$$T = 2\pi\sqrt{(m/k)}$$

where m is the mass of the glider and k is the spring constant.

Equipment: Air track, air glider with extra weights, spring, vernier calliper, tape measure and stopwatch.

Solution:
If the airtrack is inclined an angle θ and the spring is attached between the upper end of the track and the glider, it is possible to let the glider oscillate on the track. The period time for a full oscillation is measured both with and without extra weight (T_1 and T_2).

If m denotes the mass of the glider and M the mass of the extra weight, we have according to the text:

$T_1^2 = 4\pi m/k$

$$\Rightarrow T_1^2 - T_2^2 = 4\pi[m-(m+M)]/k = 4\pi M/k \Rightarrow k/M = 4\pi/(T_1^2 - T_2^2) \quad (1)$$

$T_2^2 = 4\pi(m+M)/k$

When the glider is not oscillating the gravitation will strech the spring according to Hooke's law. If x_1 and x_2 are the positions of the glider without and with extra weight:

$$\begin{cases} F_1 = kx_1 \\ F_1 = mg\sin\theta \end{cases} \quad \begin{cases} F_2 = kx_2 \\ F_2 = (m+M)g\sin\theta \end{cases} \Rightarrow$$

$\Rightarrow x_2 - x_1 = [(m+M) - m] g(\sin\theta)/k = Mg(\sin\theta)/k \Rightarrow g = (x_2 - x_1)k/(M\sin\theta)$
which combined with (1) gives

$$g = [(x_2 - x_1)/\sin\theta][4\pi/(T_1^2 - T_2^2)]$$

The stretch of the string with an extra weight $(x_2 - x_1)$, the periods T_1 and T_2 and $\sin\theta$ have to be measured.

8. (1/1984) The kinetic energy for a rolling body can be separated in translational and rotational energy. The latter is given by

$$E_{rot} = I\,\omega^2/2$$

Here I denotes the moment of inertia for the body and ω the angular velocity of rotation. The moment of inertia for a homogenous sphere is

$$I = 2M\,r^2/5$$

where M is the mass of the sphere and r its radius. Now make suitable experiments with the available equipment to get the best possible value of the gravitational constant g. Discuss sources of errors.

Equipment: Lens with a spherical concave surface, steel ball, spherometer, micrometer screw gauge, stopwatch.

Solution:
Place the lens on the table with the concave surface upwards. If the steel ball is released from the rim of the lens it will start an oscillating movement back and forth like a pendulum bob. According to the energy principle the potential energy of the ball at the rim is converted into kinetic energy at the bottom of the lens:

$$Mgh = Mv^2/2 + I\omega^2/2 \qquad \cdots \cdots \cdots (1)$$

assuming no friction. If the ball really is rolling, $\omega = v/r$ and (1) combined with the expression for the moment of inertia can be written:

$$Mgh = M\,v^2/2 + M\,v^2/5 = 7\,M\,v^2/10 \Rightarrow v^2 = (5/7) \cdot 2gh$$

The corresponding expression for a mathematical pendulum is given by

$$v_p^2 = 2gh$$

The period time for the mathematical pendulum is

$$T_p = 2\pi\sqrt{(l/g)}$$

and the corresponding period time for the rolling steel ball is

$$T = \sqrt{(7/5)T_p}$$

l for the rolling steel ball is given by l = R-r where R is the radius of curvature for the lens and r the radius of the steel ball. From this

$$T = 2\pi\sqrt{7(R-r)/5g} \quad \Rightarrow \quad g = 28\pi(R-r)/5T^2$$

R can be determined with the spherometer and r with the micrometer screw gauge. The period time for 5 or 10 oscillations can be obtained with the stopwatch.

9. (5/1987) The nowadays wellknown nuclide $^{137}_{55}Cs$ is beta-radioactive and from the disintegration scheme it is evident that the beta particles has a maximum energy of 514 keV.

A general beta spectrum, i.e. the number of beta particles (N) as a function of the energy (E_β), is shown below.

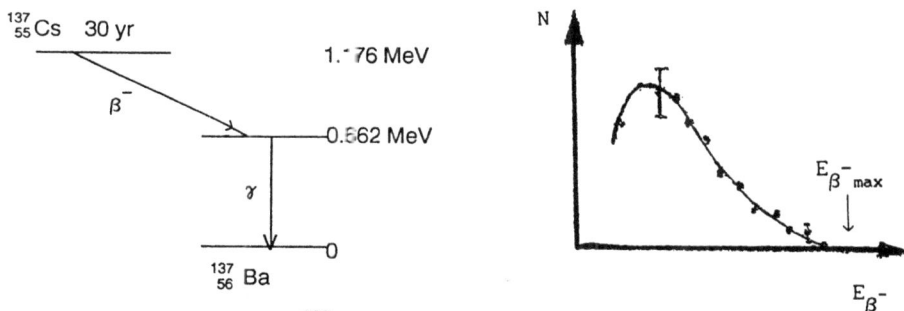

The daugther nucleus $^{137}_{56}Ba$, is not in its ground state after the disintegration, but has to emit a gamma photon. In this case, a process called internal conversion, is concurrent to the gamma emission. In that process, the energy is transferred to either a K-electron or a L-electron.

The energy in this set up, is detected with a surface barrier detector. The detector produces voltage pulses, which are proportional to the energy of the impinging particles.   These pulses are then sorted according to increasing energy in channels in a so called multichannel analyser.

Record a beta spectrum and use it to detertmine the difference in binding energy between the electrons in the K- and L-shell

Solution:
The recorded spectrum shows the continuous beta spectrum. At the high energy end are the conversion peaks from the K- and L-shell superimposed. Since the K-electrons are harder bound compared to the

L-electrons the K-peak is to the left. In addition, K-conversion is more probable and therefore the K-peak is larger. Assuming that channel 0 corresponds to 0 keV and estimating the channel number for $E_{\beta max}$ = 514 keV makes it possible to plot the energy versus channnel number. Finally the energy difference between the two conversion peaks correspond to the wanted energy difference, approx. 30 keV.

10. (5/1988) $^{201}_{81}Tl$ is a nuclide, which is used in medicine, because its atoms resembles those of potassium. It is rich of protons and distintegrates with electron capture.

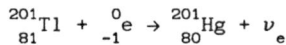

$$^{201}_{81}Tl + ^{0}_{-1}e \rightarrow ^{201}_{80}Hg + \nu_e$$

The captured electron comes in most cases from the K-shell. The vacancy in the K-shell is then filled up by an electron from some other shell, causing emission of X-rays. The daughter nucleus, $^{201}_{80}Hg$, does not necessarily go to its ground state after the electron capture, but has to emit gamma radiation to be happy. The processes are described by the disintegration scheme below.

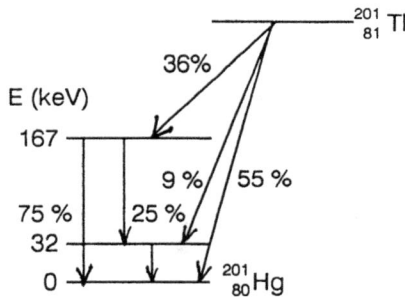

In this case the electromagnetic radiation is detected with a scintillation detector. The detector produces voltage pulses, whose height are proportional to the energy of the incoming radiation. These pulses are then sorted according to increasing energy in channels in a so called multichannel analyser.

Collect an energy spectrum and use it to calculate a lower limit for the binding energy of the K-shell electrons in mercury.

Solution:
The recorded spectrum shows the gamma peaks with energy 167 keV, 135 keV and 32 keV respectively. In addition X-ray peaks are present i.e. K_α and K_β which are not resolved and therefore looks like one peak. A plot of the gamma energies versus the channel numbers will give the energy of the X-ray peak. A value of 70 keV is reasonable.

TURKEY

The Selection and Training of Turkish Teams
for the International Physics Olimpiads

by
Prof.Dr. Ordal Demokan
Middle East Technical University, Ankara

1. Initial Selection

In Turkey, there exists a nationwide University Entrance Examination, which
is held every year around the end of June. The best 15 high schools are selected
on the basis of the performances of their graduates in this examination for that
particular year, by the Turkish Scientific and Technical Research Council. The
administrators and teachers of these Schools are then asked to select the best 4
students in Physics among the classes, which will be graduating next May. The
training group, which is composed from the staff members of the Physics Department
of the Middle East Technical University, prepares an examination for these candidates.
This examination is usually held in April and after grading by the training group,
the best 15 to 20 students are selected at this stage.

2. Training and Final Selection

These students are then invited to the Middle East Technical University at
the beginning of September and go through a training program on the fundamental
subjects of physics, such as mechanics, optics and electromagnetic theory, for 2
weeks. At the end of this period, they return to their schools and come back again
during the semester break in February. This time, they go through a training
program on thermodynamics, electrical networks and somewhat more advanced subjects,
such as modern physics and relativity for one week. During the second week, they
get acquainted with various optical and electrical devices in the laboratories of
Physics Department, carry outa few experiments and get trained on data analysis and
error calculations. At the end of these two weeks, they return to their schools
again. Sometimes they are given a number of selected problems to work on until May.

At the beginning of May, after the semester ends, they are invited
to the Middle East Technical University once again. On the second day after
their arrival, they are given a final selection examination. This year for
the first time, they will be asked to carry out an experimental task, as
well as the theoretical ones. The final team is then selected on the
basis of this exam and partly according to the opinions of the training
group, to decrease the risks of a selection based on a single exam. This
team is then hosted once more by the Middle East Technical University
for a period of two weeks, which is devoted entirely to solving and
discussing the theoretical and experimental problems of the recent
olimpiads. At the end of this period, the students return to their
families, to get prepared for the university entrance examination.

This phase constitutes the major drawback in our system, since the
students are obliged to get prepared for a very different type of
examination, concerning both the level and nature of problems, just before
the olimpiads. The efforts for providing an exemption from the university
entrance examination for our teams have unfortunately been unsuccessful
up to now. After the entrance examination, the students come back to
the Middle East Technical University just a few days before the olimpiads,
being exhausted and dequailified. These few days are usually spent for
reprocessing our team smoothly, for the type of problems they will be
given in the olimpiads.

3. Some Typical Problems asked in the Selection Examinations

1. In a cloud chamber, protons are rotating in circular orbits of radius
r_1, due to a uniform magnetic field
B, in the direction, as
shown in the figure. An unknown
particle enters the chamber in the
+x direction and after colliding
elastically with one of the protons,
scatters in a direction making an
angle of 34° with the +x direction.
The colliding proton starts
rotating around a new circular

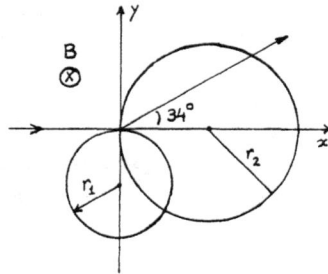

orbit with radius r_2. If $r_2 = 2r_1$, state what the unknown particle is. ($\sin 34^\circ = 0.56$, $\cos 34^\circ = 0.83$).

2. The circuit shown in the figure is laid in a uniform magnetic field B, perpendicular to its plane.
The rod of length h and mass m can slide freely on the wires. All resistances are to be neglected. The rod is given an initial velocity v_o, in the direction to the right. Describe the behavior of the velocity of the rod.

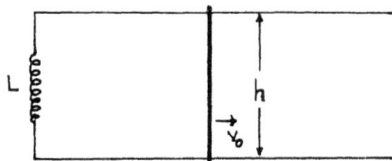

3. A thin wire is bent to form a curve, described by the parametric equations; $x = b(\theta + \sin \theta)$ and $y = b(1 - \cos \theta)$, where x and y are the horizontal and vertical axis, respectively. This wire passes through a hole in a bead, which can slide freely along the wire. Describe the motion of the bead, when released from a certain height along the wire.

4. A large and shallow, rectangular water tank is laid on the x-y plane and plane waves are generated at its upper and lower left corners, in the direction shown in the figure. The expressions for these waves can be given as,

$$\Psi_1(x,y,t) = Ae^{i(\vec{k}_1 \cdot \vec{r} - \omega t)},$$

$$\Psi_2(x,y,t) = Ae^{i(\vec{k}_2 \cdot \vec{r} - \omega t)}.$$

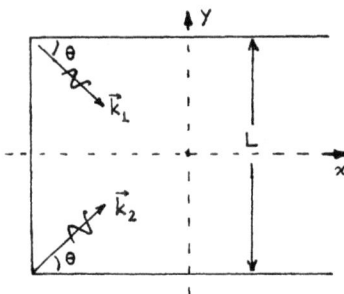

a) Show that waves propagating in the x direction but standing in the y direction are produced.

b) Find the nodes of the standing waves.

c) Find the minimum value of the frequency, required to produce undamped waves in the x direction.

Experimental Problem:

A black, cylindrical container with movable slits on its peripherial surface is given. There is one of the following items in the middle of the cylinder: a sheet of glass, a prism, a grating, a lens, a double slit or a tube filled with saturated sodium vapor. Describe an experiment to determine which item is in the cylinder.

Solutions:

1. Let the masses of the proton and unknown particle be m_p and m_x, respectively. Let the speeds before and after the collision be v_{po}, v_{xo} and v_{pf}, v_{xf}, respectively. The protons must be rotating counterclockwise in the given magnetic field. The equations for the conservation of linear momentum and energy are: $m_x v_{xo} - m_p v_{po} = m_x v_{xf} \cos 34^o$, $m_x v_{xf} \sin 34^o = m_p v_{pf}$ and $m_x v_{xo}^2 + m_p v_{po}^2 = m_x v_{xf}^2 + m_p v_{pf}^2$. Eliminating v_{xo} and v_{xf} from these equations and expressing v_{po} and v_{pf} in terms of r_1 and r_2 via the equation $v = qBr/m$, one obtains

$$r_1^2 - r_2^2 + 3r_1 r_2 = (m_x/m_p)(r_2^2 - r_1^2).$$

This equation yields $m_x = m_p$ for $r_2 = 2r_1$. Since the trajectory of the unknown particle is a straight line, it must be neutral, therefore it is a neutron.

2. Let the distance between the inductor and the rod be x. The flux through the closed circuit at any time is then Bhx. The electromotive force in the circuit is $Bh(dx/dt) = Bhv$. The current through the circuit will be $i = L^{-1} \int Bhvdt$, and the force on the rod is therefore $Bih = BhL^{-1} \int Bhvdt$. This force will be to the left for a displacement to the right. One can therefore write,

$$-m(dv/dt) = BhL^{-1} \int Bhvdt.$$

Taking the derivative of both sides with respect to t yields the equation for harmonic motion, with a frequency $Bh/2\Pi\sqrt{mL}$.

3. Let $\tan\alpha = dy/dx$. The force on the bead, tangent to the wire will be $mg\sin\alpha$. Since the forces, perpendicular to the wire cancel, this is the net force and therefore the net acceleration is $g\sin\alpha$, tangent to the wire. The component in the y direction is $-g\sin^2\alpha$. This must be equal to d^2y/dt^2. But $\sin\alpha = \tan\alpha/\sqrt{1+\tan^2\alpha}$, $dy=b\sin\theta d\theta$, $dx=b(1+\cos\theta)d\theta$, $\tan\alpha=dy/dx=\sin\theta/(1+\cos\theta)$ and $\sin^2\alpha = \sin^2\theta/(2+2\cos\theta)$. Writing $\sin^2\theta=1-\cos^2\theta$, $\cos\theta=1-(y/b)$ and substituting into the equation of motion yields

$$\frac{d^2y}{dt^2} + \frac{g}{2b}\, y = 0.$$

This is the equation for harmonic motion with period $2\pi\sqrt{2b/g}$.

4. Since the medium for both waves is the same, $\vec{k}_1 = k_x\hat{x}-k_y\hat{y}$ and $\vec{k}_2=k_x\hat{x}+k_y\hat{y}$. Writing $\vec{r}=x\hat{x}+y\hat{y}$, using the relation $\exp(ix) = \cos x + i\sin x$ and common trigonometric identities, the total wave function can be obtained as,

$$\psi = \psi_1 + \psi_2 = 2A\cos k_y y . e^{i(k_x x-\omega t)} .$$

This is the expression for a wave propagating in the x direction and standing in the y direction. The position of the nodes can be obtained by letting $k_y y = (2n + 1)\pi/2$, where $n=0,1,2 \ldots$. If the speed of the waves is v and the frequency is ω, $k= \omega/v$ and $k_y= k \sin\theta = \omega \sin\theta/v$, yielding

$$y = (2n + 1)\ \pi v/2\omega \sin\theta.$$

For undamped waves in the x direction, k_x must be real. Since $k^2 = k_x^2+k_y^2$, one can write $\omega^2 = v^2(k_x^2+k_y^2)$. The minimum value for the frequency is then $\omega_c=v\, k_{ymin}$. If the channel width (in the y direction) is L, maximum wavelength is 2L and therefore $k_{ymin} =\pi/L$. This gives $\omega_c = \pi v/L$.

Experimental problem:

A collimated light beam from either a Na or Hg source is sent into the cylinder through one slit and observed through the other. For the case of a sheet of glass, the incident and emerging beams will be parallel. A prism will deflect the incident beam and a lens will focus it, which can be observed by moving a screen at the emerging side. In the case of a grating, a symmetrically lined spectrum around the fundamental band will be observed. For the double slit, both interference and diffraction effects will be produced. Nothing will be observed in the case of saturated Na vapor, when Na lamp is used.

UNITED KINGDOM

THE BRITISH PHYSICS OLYMPIAD

Dr. C. Isenberg
University of Kent at Canterbury, CT2 7NR.

and

Mr. Guy Bagnall
Harrow School, Middlesex, HA1 3JE.

THE ORGANIZATION OF BRITISH COMPETITION

Each year all schools in Britain receive a Test Paper which teachers are encouraged to set to the final year students. This is not an overly difficult paper. Teachers mark these papers and encourage those students who obtain marks over 70% to enter the more challenging, annual, British Physics Olympiad Examination which is held in February. Last year nearly 500 students entered this examination.

The students take the 3 hour theoretical examination at their schools. Teachers forward the scripts to the central Olympiad Office, at the University of Kent, to be marked. The marks are graded into classes. The top twenty marks provide the Gold medal award winners, the next 50 the Silver medal award winners, the subsequent 125 the Bronze medal awards, followed by 150 Nickel medal awards. The remainder are in the Commendation class. All medal award winners receive prizes at a ceremony at the Institute of Physics in London.

The Gold award winners are invited to London to take a challenging $3\frac{1}{2}$ hour experimental examination plus a one hour Advanced Theoretical Examination. The five selected British Physics Olympiad Team members are those with the highest total marks in all the examinations. In past years we have not held a training session for the British Physics Olympiad Team prior to the

the International Physics Olympiad. However, this year we held a weekend training session. During the training sessions the students were tutored in areas they do not cover at school, such as A.C. theory, thermodynamics, geometrical and physical optics.

In the following two sections examples are given of: (i) questions and solutions from past British Theoretical Examination Papers; (ii) two previous Experimental Examination Papers.

Some Questions and Solutions from British Physics Olympiad Examinations.

Question 1.

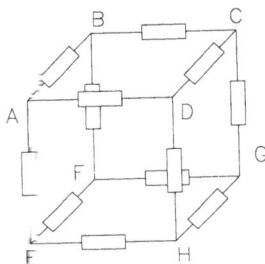

The diagram shows a cube ABCDEFGH with resistors, each of resistance r, joined by conductors along the edges of the cube. Calculate the total resistance between:

(i)   AG

(ii)   AD

If the resistors between BF, CG and DH are short circuited calculate the resistance between AG.

Solution.

(i)

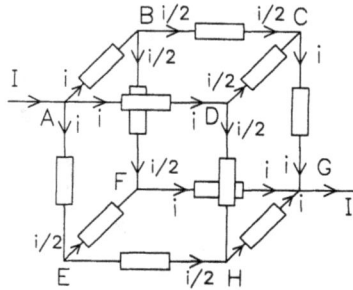

By symmetry three arms of cube at A have equal currents i. Symmetry w.r.t. reversal of PD across AG gives three equal currents i at G. By symmetry, Kirchoff's first law at B,D,C,D,E,F and H requires current i to split into two equal currents i/2

Resistance R_{AG}, betwen A and G, is given by

$$R_{AG} = \frac{V_{AG}}{I} = \frac{V_{AD} + V_{DC} + V_{CG}}{3i} = \frac{\frac{5}{2} \, i \, r}{3i}$$

$$= \frac{5}{6} \, r$$

(ii)

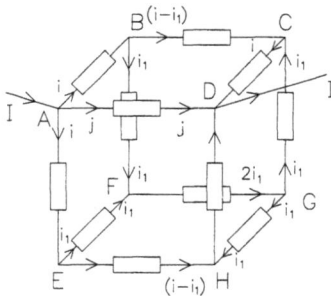

By symmetry and reversibility arms, AB, AE, CD and HD have same current i. Let AD have current j. By symmetry, current i at nodes B, E, H and C will split into i_1 and $i-i_1$ as indicated. Arm FG will have current $2i_1$.

In loop EFGH: $i = 5i_1$ (1)
In loop ABCD: $j = 3i - i_1 = 14i_1$ from (1)   (2)

Resistance R_{AD}, between A and D, is given by

$$R_{AD} = \frac{jr}{I} = \frac{jr}{2i + j} = \frac{14i_1 r}{24i_1}$$

$$= \frac{7}{12} r$$

————————

(iii)

(a)

(b)

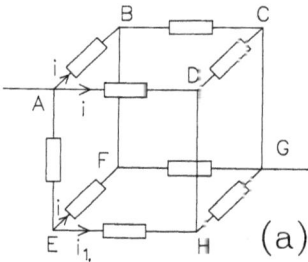

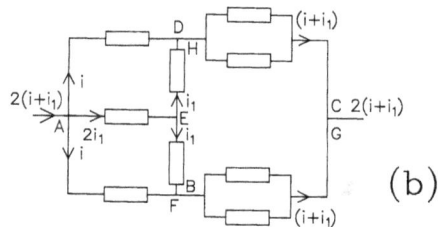

Shorting BF, CG and DH, the circuit retains, by symmetry, equal currents i in AE and AD and equal currents i_1, say, in EH and EF. Kirchoff's 1st law at E gives current 2i in AE. Simplifying the circuit (a) gives circuit diagram (b). Let us now consider circuit (b).

Kirchoff's first law: Current in AE $= 2i$

Resistance between DG and BG is equal to $\left(\dfrac{1}{r} + \dfrac{1}{r}\right)^{-1} = \dfrac{r}{2}$

and have currents $(i + i_1)$

Kirchoff's second law in AEF gives $\qquad i = 3i_1 \qquad (1)$

$$\therefore\ P_{AG}\ =\ \frac{ir + \dfrac{r}{2}\,(i + i_1)}{2(i + i_1)}\ =\ \left(\frac{5i_1}{(8)i_1}\right)\ r\ \text{ from (1)}$$

$$=\frac{5}{8}\,r$$

─────

Alternative Method of Solution

(i)   B, D and E are, by symmetry, at the same potential so can be connected by a conductor without changing the currents in the arms of the circuit. A similar result is true for C, H and F. So one obtains the equivalent circuit (all resistors being r):

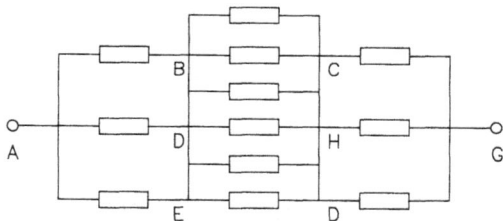

Adding the parallel sets of 3, 6 and 3 resistors ,the circuit reduces to:

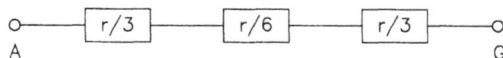

Total resistance between AG $\ =\ \left(\dfrac{r}{3}\ +\ \dfrac{r}{6}\ +\ \dfrac{r}{3}\right) = \dfrac{5}{6}\,r$

─────────────────────────────────────

(ii)   B and E are, by symmetry, at the same potential so can be
       joined by a conductor. A similar result holds for C and H.
       The circuit thus reduces (all resistors being r) to:

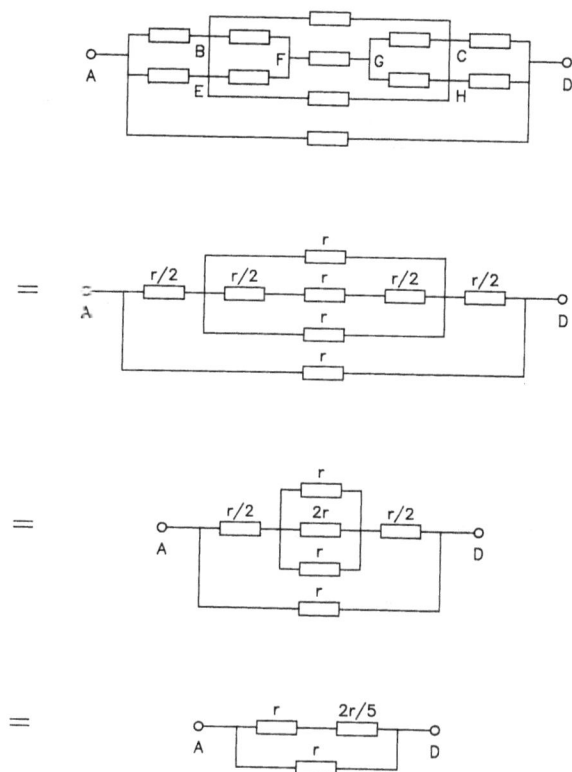

$$\text{Total resistance between A and D} \quad = \quad \left(\frac{5}{7\,r} + \frac{1}{r}\right)^{-1} \quad = \quad \frac{7}{12}\,r$$

(iii)   B and F are at the same potential and D and H are also. By symmetry all these four points have the same potential. The equivalent circuit between A and G:

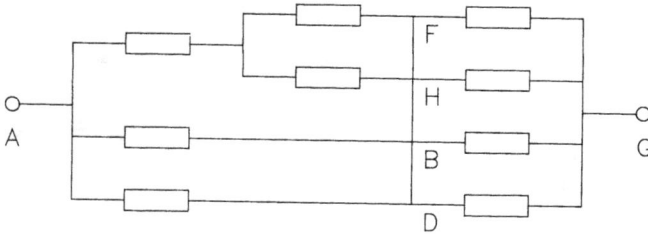

Adding resistances in series and parallel gives the equivalent circuit:

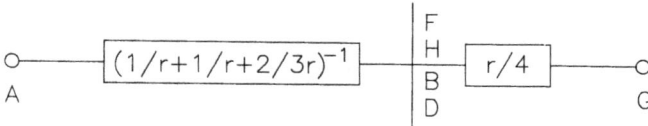

Total resistance between AG $= \left(\dfrac{1}{r} + \dfrac{1}{r} + \dfrac{2}{3r}\right)^{-1} + \dfrac{r}{4} = \dfrac{3}{8}r + \dfrac{r}{4} = \dfrac{5}{8}r$

Question 2.

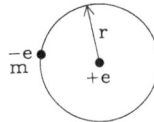

(a) Rutherford postulated that the hydrogen atom had a negatively charged electron of mass m which is constrained to move, with angular velocity ω, in a circular orbit by the electrostatic attraction of the positively charged, stationary, nucleus. Show that the total orbital energy E of the electron, provided the atom is in a non-radiating state is given by

$$E = - \frac{e^2}{8\pi\varepsilon_0 r} \, ,$$

where r is the orbital radius and ε_0 is the permittivity of free space.

(b) The results obtained at the end of the nineteenth century from spectroscopic experiments can be represented by

$$E = - \frac{me^4}{32\pi^2\varepsilon_0^2 H^2} \frac{1}{n^2}$$

where H is a constant and n is an integer. Deduce from these results Bohr's postulate that, if the atom does not radiate electromagnetic energy, the angular momentum (or moment of momentum) of the electron about the nucleus can have only discrete values nH.

(c)   Assuming Planck's hypothesis for the quantization of radiation, show that if a single quantum is emitted when an electron makes a transition from an orbit with number n to one with quantum number (n-1) then for large values of quantum n the radiation frequency v will be classical and related to E by

$$\nu = c\,|\,E\,|^{3/2}, \text{ where c is a constant.}$$

(d)   Classical physics requires that the frequency emission of radiation be idential to the frequency of rotation of the electron around the nucleus. If, in the limit of large n, this classical result is assumed, deduce that the Planck constant h is equal to $2\pi H$.

Solution.

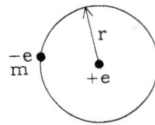

$$E = \frac{m}{2}(r\omega)^2 - \frac{e^2}{4\pi\varepsilon_0 r} \qquad (1) \quad (E = KE + PE).$$

Now force acting on electron

$$\frac{e^2}{4\pi\varepsilon_0 r^2} = mr\omega^2 \qquad (2) \quad \text{for circular motion.}$$

Substituting into (1) for ω^2

$$E = \frac{1}{2}\frac{e^2}{4\pi\varepsilon_0 r} - \frac{e^2}{4\pi\varepsilon_0 r} \quad .$$

$\therefore \quad E = -\frac{1}{2}\frac{e^2}{4\pi\varepsilon_0 r} = -\frac{e^2}{8\pi\varepsilon_0 r} \qquad (3) \quad .$

Angular momentum $L = mr^2\omega \qquad (3a)$.

$\therefore \quad L^2 = m^2 r^3 (r\,\omega^2)$

$$= mr\frac{e^2}{4\pi\varepsilon_0} = \frac{me^2 r}{4\omega\varepsilon_0} \qquad (4) \text{ from } (2) \quad .$$

$\therefore \quad$ Substituting (4) into (3)

$$E = -\frac{me^4}{32\pi^2\varepsilon_0^2 L^2} \qquad\qquad \text{as } \frac{1}{r} = \frac{me^2}{4\pi\varepsilon_0 L^2}$$

$$= -\frac{\omega^4 m}{32\pi^2\varepsilon_0^2 H^2}\frac{1}{n^2} \qquad (5) \text{ spectroscopic result} \quad .$$

Hence $\underline{L = nH} \qquad\qquad (5a)$

Planck: $h\nu = E_n - E_{n-1}$

$$= -\frac{me^4}{32\pi^2\varepsilon_0^2 H^2}\left[\frac{1}{n^2} - \frac{1}{(n-1)^2}\right]$$

$$= \frac{me^4}{32\pi^2\varepsilon_0^2 H^2}\frac{2n-1}{(n-1)^2 n^2}$$

$$= \frac{me^4}{32\pi^2\varepsilon_0^2 H^2}\frac{2}{n^3} \qquad\qquad (6) \text{ for large } n$$

∴   from (5)

$$E = - \left(\frac{me^4}{32\pi^2\varepsilon_0^2H^2}\right)\left(\frac{32\pi^2\varepsilon_0^2H^2h\nu}{2me^4}\right)^{2/3} = - \left(\frac{me^4}{32\pi^2\varepsilon_0^2H^2}\right)^{1/3}\left(\frac{h}{2}\right)^{2/3}\nu^{2/3}$$

∴ $\nu \propto |E|^{3/2}$.

Now from (6)

$$\nu = \frac{m^4e}{16\pi^2\varepsilon_0^2H^2h}\left(\frac{1}{n^3}\right). \qquad (7)$$

Now from, (3a) and (5a)

$$mr^2\omega = nH \quad \text{where } \omega \text{ is ang.freq. of rotation.}$$

If ν' frequency.of revolution,.

$$\nu' = \left(\frac{\omega}{2\pi}\right) = \frac{nH}{(2\pi)m}\frac{1}{r^2}$$

$$= \frac{nH}{(2\pi)m}\frac{(8\omega\varepsilon_0)^2E^2}{e^4} \quad \text{from (3)}$$

$$= \frac{nH4(4\pi\varepsilon_0)^2}{(2\pi)me^4}\left(\frac{e^4m}{32\pi^2\varepsilon_0^2H^2}\right)^2\frac{1}{n^4} \quad \text{from (5)}$$

$$= \frac{e^4m}{(4\pi\varepsilon_0)^22\pi H^3}\frac{1}{n^3} \qquad (8)$$

Equating (7) and (8)

$$\underline{h = 2\pi H}$$

Question 3.

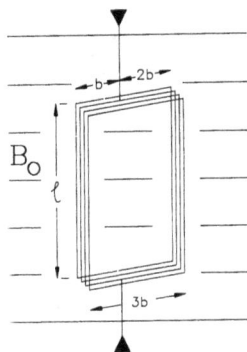

(a) A rectangular coil of wire with vertical length l and horizontal width 3b has N turns and forms a closed circuit. It has an angular velocity ω, at time t, about the vertical axis a distance b from one vertical side of the coil, in a constant horizontal flux density B_0, as indicated in the diagram. Show that the work done in rotating the coil against the action of the magnetic field, in a small time interval, is equal to the energy dissipated by the induced emf in the coil.

(b) The coil has a resistance R and a self inductance L. Write down the equation satisfied by the current I in the coil at time t if at t = O the plane of the coil is parallel to the field. When ω is constant, verify by substituting into the equation, that the equation has a solution of the form

$$I = I_0 \cos(\omega t + \phi)$$

where I_0 and ϕ are constants. Deduce that the amplitude I_0 depends inversely on the square root of $(R^2 + \omega^2 L^2)$, and determine the phase ϕ.

(c) Calculate the total energy dissipated during one revolution of the coil in (i) the resistance and (ii) the inductance in terms of I_0, R, L and ω.

Solution

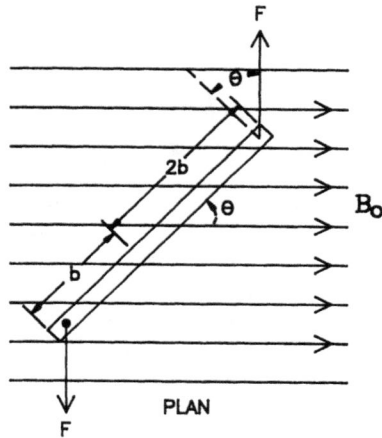

Let I be current in wire coil at time t when inclined at angle θ to field B_0.

Force F on coil given by

$$F = IlB_0N \qquad (1)$$

in direction perpendicular to B_0 as indicated.
Work done during rotation $d\theta$

$$
\begin{aligned}
dW &= F\cos\theta(2b)d\theta + F\cos\theta(b)d\theta \\
&= 3Fb\cos\theta\, d\theta \\
&= 3Ilb_0\cos\theta\, d\theta \qquad \text{From (1)}
\end{aligned}
$$

$$\therefore \qquad dW = 3I\omega lbNB_0\cos\theta\, dt \quad (1) \quad \text{as } \frac{d\theta}{dt} = \omega(t)$$

Rate of working $= 3I\omega lNB_0\cos\theta$ (note $\theta = \int^t \omega dt$)

Magnitude e.m.f. produced in coil $\varepsilon = \dfrac{d\phi}{dt}$ where $\phi = $ flux due to B_0

$$\varepsilon = \frac{d}{dt} (B_0 \sin\theta) 3Nlb$$

$$\therefore \quad \varepsilon = 3\omega lb NB_0 \cos\theta \quad \text{as} \quad \frac{d\theta}{dt} = \omega(t) \qquad (2)$$

Energy dissipated in coil in time dt,

$$dW' = \varepsilon I dt$$

$$\therefore \quad \underline{dW' = 3\omega l B_0 \cos\theta \ dt} \qquad \text{from } (2)$$

$$\therefore \quad \underline{dW = dW'}$$

From (2)

$$\varepsilon = RI + L\frac{dI}{dt}$$

i.e. $\quad \varepsilon_0 \cos\theta = RI + L\frac{dI}{dt}$

where $\varepsilon_0 = (3\omega lb NB_0)$ from (2)

$$\varepsilon_0 \cos\theta = RI + L\frac{dI}{dt}$$

Substituting $I = I_0 \cos(\omega t + \phi)$ for the case of constant ω, in which $\theta = \omega t$,

$$\varepsilon_0 \cos\omega t = RI_0 \cos(\omega t + \phi) - LI_0\omega \sin(\omega t + \phi)$$

$$= I_0\sqrt{R^2 + \omega^2 L^2} \left[\frac{R}{\sqrt{R^2 + \omega^2 L^2}} \cos(\omega t + \phi) - \frac{\omega L}{\sqrt{R^2 \omega^2 L^2}} \sin(\omega t + \phi) \right]$$

$$= I_0\sqrt{R^2 + \omega^2 L^2} \ [\cos\alpha \ \cos(\omega t + \phi) - \sin\alpha \ \sin(\omega t + \phi)]$$

$$= I_0\sqrt{R^2 + \omega^2 L^2} \ \cos(\omega t + \phi + \alpha)$$

Comparing LHS and RHS of this equation

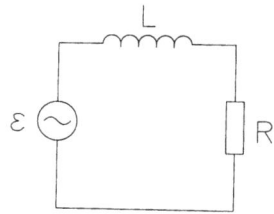

$$I_0 = \frac{\varepsilon_0}{\sqrt{R^2 + \omega^2 L^2}} \text{ and } \phi = -\alpha = -\tan^{-1}\left(\frac{\omega L}{R}\right)$$

In one revolution of coil, energy dissipated in resistance

$$= \int_0^{2\omega/\omega} RI(I)dt = I_0^2 R \int_0^{2\pi/\omega} \cos^2(\omega t + \phi)dt$$

$$= I_0^2 R \int_0^{2\pi/\omega} \frac{1 + \cos 2(\omega t + \phi)}{2} dt$$

$$= \frac{1}{2} I_0^2 R \frac{2\omega}{\omega} = \pi I_0^2 R/\omega$$

In one revolution of coil energy dissipated in inductance

$$= \int_0^{2\pi/\omega} I L \frac{dI}{dt} dt = L \int_0^{2\pi/\omega} I \, dI = L[I^2/2]_0^{2\pi/\omega}$$

$$= 0 \text{ as } I = I_0 \cos(\omega t + \phi)$$

EXPERIMENTAL EXAMINATIONS

1988 BRITISH PHYSICS OLYMPIAD

INTRODUCTION

This is a 3 hour experiment the aim of which is to estimate, from measurements at the temperature of liquid nitrogen, the magnetic moment and current in a high temperature superconductor (HTS) using a simple model.

When a current is generated in the HTS, the superconductor contains a closed circulating current I enclosing a planar area A. The magnetic moment M due to the current is given by

$$M = IA,$$

The superconductor behaves like a current I circulating in a wire enclosing a planar area A. A may be approximated by the cross sectional area of the superconductor.

PROCEDURE

Suspend the HTS from a clamp and stand, using the cotton provided. It should lie symmetrically between the poles of the magnet at the end of the pole pieces, as indicated in Figures 1 and 2. The horizontal line symmetrically through the poles of the magnet, indicated in Figure 1, is designated the x-axis. The z-axis is vertical and the y-axis is horizontal. The magnet rests on a lab jack which can be raised and lowered.

The specimen should be placed in the liquid nitrogen contained in a small polystyrene cup, (This can be refilled from a Dewar flask as necessary). It should be left to cool for at least 30 secs, afterwards it can be pulled out by the cotton and put in the magnetic field. It will lose its superconductivity as the specimen warms up so the position it takes up must be noted quite quickly.

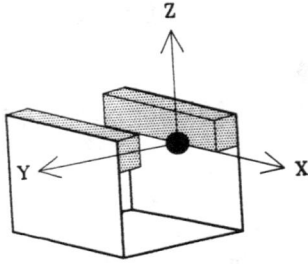

FIGURE 1

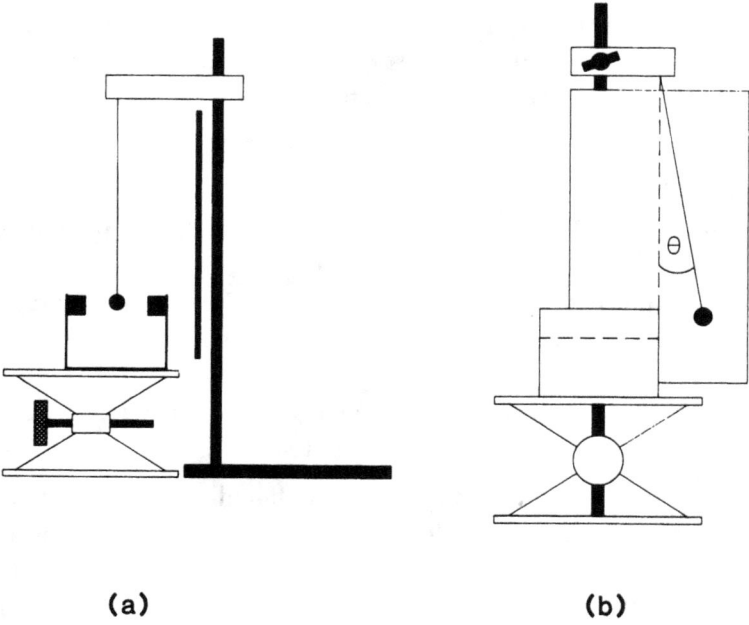

(a)   **(b)**

FIGURE 2

The force experienced will be related to the angle θ the cotton makes with the vertical. It may be necessary to repeat the cooling process if the superconductor does not produce the 'θ' displacement shown in figure 2a.

(N.B. The polystyrene cup should be no more than one third full of liquid nitrogen. The cup can then be handed safely by grasping its rim.)

INSTRUCTIONS

1. A Hall probe is provided which must be calibrated using the "infinite" solenoid provided.

    The formula for the magnetic field in an 'infinite' solenoid is $B = 4\pi\ 10^{-7}I\ N/L$ where N is the number of turns of the solenoid, L its length and i is the current in the solenoid.

    A graph of the variation of magnetic field strength B against Hall voltage should be determined. The Hall probe must be run from the 5V power supply.

2. The force in the x-direction, F, on the circulating current I, produced in the superconductor and enclosing an area A perpendicular to the x-axis, which causes the displacement of the HTS in the magnetic field is,

    $$F = M\ dB/dx$$

    where M = IA. All other magnetic forces can be neglected.

3. Measure the linear displacement of the superconductor and determine θ.

4. Draw a graph B against x and determine the value of the gradient dB/dx at the position of the displaced specimen.

5. Measure the mass of the specimen using the electronic balance.

6. From the equilibrium of the forces acting on the HTS determine the value of M. Make an estimate of the area of the specimen perpendicular to the x axis. Hence obtained an estimate for I.

7. Is the magnetic field configuration described in this experiment optimal for demonstrating the force on the HTS?

8. Comment on the accuracy and validity of the experiment.

9. Using a diagram of this model prove the result

$$F = IA \, dB/dx.$$

APPARATUS FOR EXPERIMENT

1. Lab jack.
2. Voltmetre (200 mV)
3. Superconductor on fine cotton
4. Ceramic magnet on yoke
5. Optical pins
6. Scissors
7. Clamp stand and poles
8. Cardboard
9. Graph Paper
10. Hall probe attached to rod.
11. Solenoid
12. 12V light source
13. 5V stabilised power supply
14. Metre rule
15. 0-5A ammeter DC
16. 8V DC max 5A. (This is the bench source for the solenoid)
17. 16 ohm rheostat
18. Polystyrene cup.

Access to: (i)   liquid nitrogen
                   (ii)   10 mg electronic balance

1989 BRITISH PHYSICS OLYMPIAD

This experiments lasts for $3\frac{1}{2}$ hours. It is necessary to plan your programme of work and carry it out efficiently.

CARE

Avoid looking directly into the quartz iodine lamp (QI).

Whenever possible shield the lamp from the eyes

There is a remote possibility that the lamp could fail by explosion so the shield should be kept on whilst the light is on.

Data

You may find the following helpful

Velocity of light $c = 3.00 \times 10^8$ ms^{-1}

Specific heat capacity of copper 380 J K^{-1} Kg^{-1}

Apparatus

QI lamp 50W.

Digital Voltmeter.

Thermocouple attached to a small piece of blackened copper shielded by a polystyrene cone and attached to an amplifier in a small white box.

5B stabilised power supply for the amplifier output, reading 0.1V per°C. (Assume correctly calibrated).

Stop watch.

Variable power supply for the QI lamp (AC transformer or DC power).

Ammeter and voltmeter (AC for the ac driven lamps).

Copper calorimeter or beaker with ice.

Glass block transmission 400 - 1200 nm.

IR red filter transmission 750 - 1000 nm.

Graph paper.

Stefan print outs.

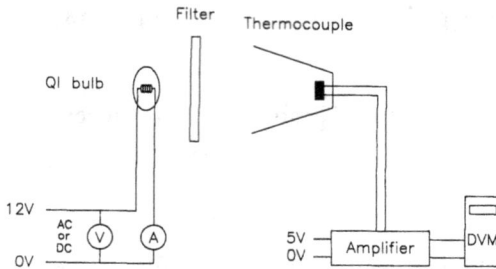

THEORY

The total electromagnetic radiation energy per unit time emmitted, for all waveleghts, by a black body (one that absorbs all radiation falling on it) is given by

$$E = \sigma A T^4$$

Where σ is the Stefan Boltzmann constant and A is the area of the emmitting surface. The QI lamp is a good approximation to a black body. Curves are provided showing how the power is distributed over the range of wavelengths. The distribution function $g(\lambda)$ is plotted against λ. The integral of g over all wavelengths is E.

AIM

This experiment measures the total radiation from a QI lamp at different running temperatures. Investigation using two filters are carried out and simple calorimetric measurements are made.

PROCEDURE

The small black disc attached to the thermocouple should be placed 20 cm from the QI lamp. This distance should not be altered during the course of the experiment.

A)   Plot graphs of the temperature of the black disc against time for a constant lamp temperature for no filter, for the IR filet, and the Glass block.   Repeat for at least four different powers of the lamp.   Since the lamp temperature is liable to fluctuate after it is switched on it is wise to monitor the current to make sure it has settled down.   The thermocouple may be pre-cooled by inserting the ice cold container between it and the light.   Determine graphically the rate of rise of temperature at room temperature.   (Do not go too far below and above room temperature).   Note the Electrical Power used by the lamp.

B)   Plot the rate of rise of temperature of the copper AT ROOM TEMPERATURE against the power for the three situations

    (i)   No filter present

    (ii)   Glass block present

    (iii)   IR filter present

What is the importance of using the temperature gradient at room temperature?

C)   One can show by making appropriate assumptions that the current I and the voltage V of the lamp are related by

$$V^3 = C \times I^5$$

where C is a constant.

Use your measurements to verify this relationship experimentally.   Derive this relationship theoretically stating the assumptions made.

D)   It is possible to determine the temperature of the lamp by observing that different proportions of the energy are emitted at different wavelengths for different filament temperatures .   Use the given Stefan-Boltzmann curves to determine the temperature of the filament.

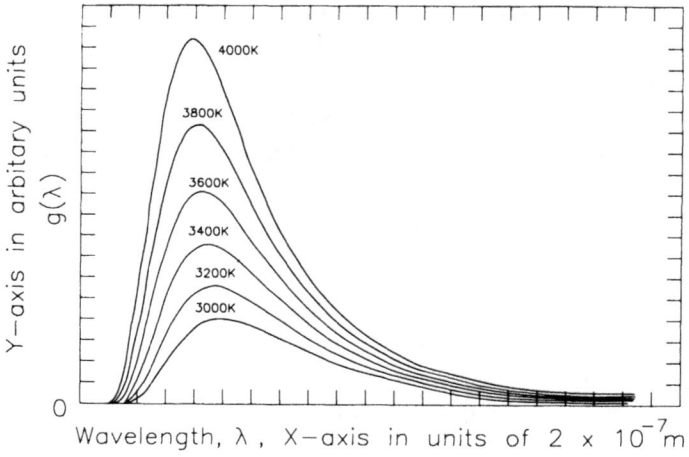

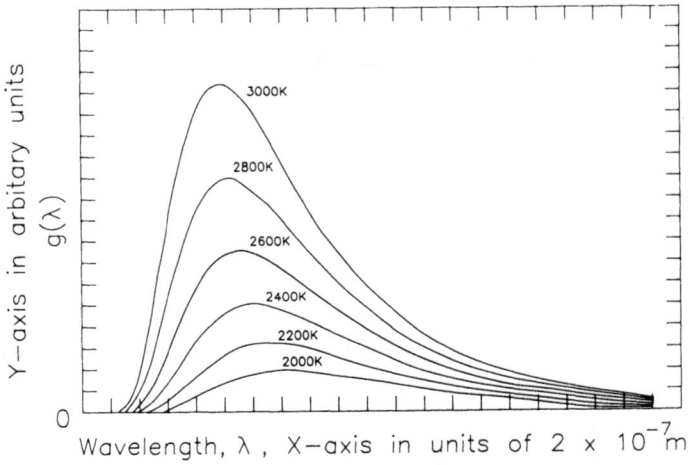